BASIC
HUMAN PHYSIOLOGY

OUTLINES & ESSENTIAL CONCEPTS

LEARNING OBJECTIVES

OUTLINES FOR STUDY

REVIEW QUESTIONS

THIRD EDITION

Richard Pflanzer, Ph.D.

**INDIANA UNIVERSITY - PURDUE UNIVERSITY
INDIANAPOLIS**

KENDALL/HUNT PUBLISHING COMPANY
4050 Westmark Drive   Dubuque, Iowa 52002

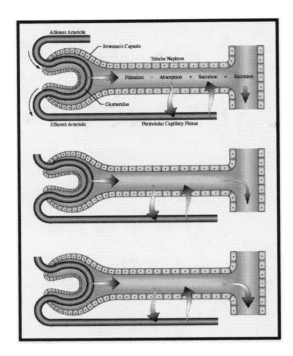

Cover image provided by Richard Pflanzer

Copyright © 1997, 2001, and 2004 by Richard Pflanzer

ISBN 13: 978-0-7575-0822-6

ISBN 10: 0-7575-0822-7

Printed in the United States of America
10 9 8 7 6 5 4 3 .

CONTENTS

PREFACE

This coil-bound, perforated series of physiology learning objectives, outlines, tables, drawings, note pages, and review questions is intended to help students as they learn basic concepts of human physiology. Please bring this book to class and use it for reference and for note-taking. A blank note page accompanies each page of outline and the note page contains an area where questions that arise during your reading or listening to a related oral presentation may be jotted down to be asked at a later date.

Use the outline and the course schedule to guide your textbook reading prior to attending the class. Prior reading will enable you to listen more critically, take notes more effectively, derive a greater benefit from instruction, and make it easier for you to meet the goals stated as learning objectives.

Learning objectives for concepts designated in the outline are listed in numerical order and precede each outline. The objectives are goals to be attained as you learn basic concepts of human physiology contained within each outline. Review the objectives just prior to attending the related lecture. The review will help you to listen critically and focus your attention on explanations that will assist you in meeting the learning objectives. After listening to the lecture and reading assigned related materials, you should be able to successfully answer/respond to each of the listed objectives for that lecture. Examination questions will be derived from the learning objectives.

Review questions have been appended at the end of each outline. These questions are helpful when properly used in a program of disciplined study and self-testing. If used improperly (memorizing), the questions are of little value, and in fact can be detrimental. Use the review questions as a self-test of your knowledge after you have read the assigned material, and have reviewed your class notes. Questions that you don't understand or mark incorrectly will point to deficiencies in your knowledge base and/or study techniques. With each question you should be able to explain why the associated answers are either correct or incorrect. This approach broadens your knowledge base regarding the concepts covered by the questions.

READING ASSIGNMENTS

TEXTBOOK :_____

AUTHORS :_____

PUBLISHER:_____ EDITION:_____PUB. DATE:_____

OUTLINE	ASSIGNMENT	OUTLINE	ASSIGNMENT
1		15	
2		16	
3		17	
4		18	
5		19	
6		20	
7		21	
8		22	
9		23	
10		24	
11		25	
12		26	
13		27	
14		28	

SUGGESTIONS

Not all of the material covered in the text reading assignments will be covered in class. You will be responsible for reading the assigned material. Examination questions will be derived from lecture, text, and laboratory presentations. Relative to the assigned textbook reading, I offer the following suggestions for making the most efficient use of your time:

1. Read the learning objectives and lecture outline in advance of the lecture presentation.
2. Based on the learning objectives and the outline, read the segments in the text related to the lecture material prior to attending lecture.
3. Review notes and text material soon after the lecture. If necessary, rewrite your lecture notes into a more legible form.
4. Complete the remainder of the assigned reading prior to the examination.
5. Obtain answers to any questions that may arise, from either textbook or lecture material, during the portion of the lecture period set aside for questions and answers, or by scheduling an office appointment.
6. Use the review questions at the end of the outlines and at the back of this book, and at the end of the textbook chapters as a self-test of your knowledge.

NOTES

DATE: _____   *OUTLINE #_____*

QUESTIONS

1._____
2._____
3._____
4._____
5._____

OUTLINE 1

LEARNING OBJECTIVES (√)

After reading the assigned pages in the textbook, and/or reading and performing related laboratory exercises, and listening to lecture and/or laboratory presentations, you should be able to:

study of nature — study of living function

☒1. Explain the meaning of the word physiology based on its Latin/Greek derivatives and give a modern definition of physiology.

☒2. Explain the meaning of the word homeostasis based on its Latin/Greek derivatives, give a modern definition of the word, and explain its importance as a theme in physiology.

☒3. Define and give an example of a negative feedback control system and explain the importance of negative feedback control systems in homeostasis. *temp*

☒4. Define and give an example of a positive feedback control system and explain why positive feedback control systems are not common in the human body. *Labor*

☒5. Outline the general structural and functional organization of the human body. *cell - tissue - organ system*

☐6. Define and give an example of cell "organelle" and cell "inclusion".

☐7. Identify each of the following organelles on a cell diagram and list the major functions of each organelle: nucleus, lysosome, Golgi apparatus, mitochondrion, smooth and rough endoplasmic reticulum, microtubules, microfilaments, and centrosome.

☐8. Diagram the fluid-mosaic model of the plasma membrane, including phospholipids, glycolipids, peripheral proteins, integral proteins, glycoproteins and cholesterol.

☐9. List three functions of a plasma membrane's integral proteins.

☐10. Explain the difference between ciliary movement and ameboid movement, and give an example of each in the body.

☐11. Define the cell property of irritability and the terms positive stimulus and negative stimulus.

☐12. Explain the meaning of respiration and circulation as applied to general cell activities.

☐13. Define anabolism, catabolism, metabolism, and their interrelation.

☐14. Explain how the cellular processes of secretion and excretion are similar yet different.

☐15. List components of the cell cycle, and outline the phases of mitosis.

, Pos. feedback - uterus - can rupture - doesn't stop til done

, Neg. feed back - heater - thermostat

, Controlled Elements - tell what to do - constant

, Regulated Elements - assume its okay - tell when not

, cells working together - throughout whole body

cell membrane = plasma membrane

head - phospho - tail - lipids

Passive & Energy

QUESTIONS

1. _____
2. _____
3. _____
4. _____
5. _____

OUTLINE 1

LECTURER:_____   NOTES_____   DATE:_____

I.   **INTRODUCTION**
    A.   **Definitions**
        1.   Physiology
            a.   Literal (from Greek/Latin)

study of nature

            b.   Modern –

study of living functions

        2.   Homeostasis
            a.   Literal (from Greek/Latin)

Remain the same

            b.   Modern

body's Response to maintain internal stability

    B.   **Control Systems**
        1.   Negative feedback
        Definition: *works to Restore normal Value*

Diagram: control of blood glucose

        2.   Positive feedback
        Definition: *set point needs to be reached destabilizes systems*

Diagram: gastric secretion

4

NOTES

QUESTIONS

1. _____
2. _____
3. _____
4. _____
5. _____

C. **General Body Organization**
1. Cells (neuron, skeletal muscle, red blood cell, etc.)
2. Tissues (epithelial, muscular, nervous, connective)
3. Organs (stomach, intestine, liver, heart, lung, etc.)
4. Systems (circulatory, respiratory, nervous, endocrine, etc.)

II. **CELL STRUCTURE AND FUNCTION**
A. General Classifications: Size, Shape, Functions

B. Organelles
Definition: *organs of the cell*

Example: *ER, golgi app., mitrocondria*

C. Inclusions
Definition: *particles temporarily in cytosol*

Example:

D. Organelles: Structure & Function
1. Membraneous organelles
a. <u>Plasma membrane</u>
Diagram: fluid mosaic model *pg 110*

functions:

b. <u>Nucleus</u>
Diagram: structure of the nucleus

functions:

NOTES

DATE : _____

OUTLINE #_____

QUESTIONS

1._____
2._____
3._____
4._____
5._____

GENERALIZED ANIMAL CELL

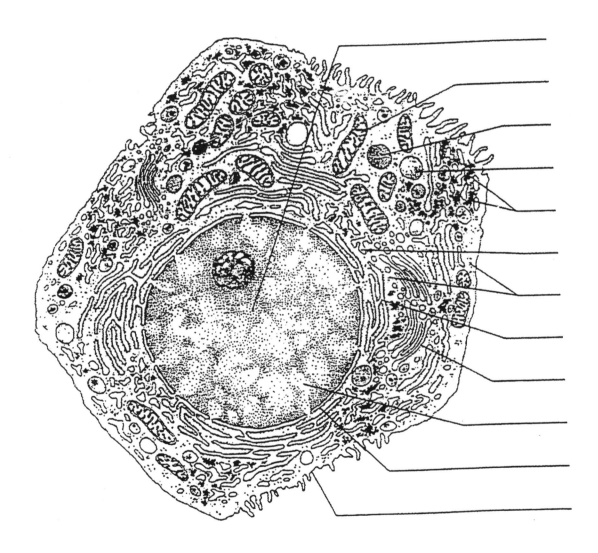

NOTES

DATE : ________

*OUTLINE #*________

QUESTIONS

1.____
2.____
3.____
4.____
5.____

c.   Rough endoplasmic reticulum
        Diagram: rough ER structure

        functions:

d.   Smooth endoplasmic reticulum
        Diagram: smooth ER structure

        functions

e.   Mitochondrion
        Diagram: mitochondriom structure

        functions

f.   Lysosome
        Diagram: lysosome structure

        functions:

g.   Golgi apparatus
        Diagram: Golgi structure

        functions:

2.   Non-membranous organelles
    a.   Microtubules
            Diagram: microtubule structure

            functions:

NOTES

DATE : _____   *OUTLINE #_____*

QUESTIONS

1._____

2._____

3._____

4._____

5._____

    b.   Microfilaments
          Diagram: microfilament structure

          functions:

E.   General Cell Activities
      1.   Movement
            a.   Ciliary
                  Diagram: ciliated cells and movement of cilia

            b.   Ameboid
                  Diagram: leukocyte movement

      2.   Irritability and Stimulus Response
            a.   Positive stimulus (excitatory) (+)

            b.   Negative stimulus (inhibitory) (–)

      3.   Respiration (O_2 consumption and CO_2 production)

      4.   Circulation

      5.   Metabolism
            a.   Definition

            b.   Anabolism

            c.   Catabolism

      6.   Excretion vs. secretion

      7.   Reproduction: cell cycle and mitosis

NOTES

QUESTIONS

1. ⑧ Psuedopodium - ameBod
2. ⑪
3. ⑮ Ectoprotiens - outer proteins
4. _____
5. _____

Must Do all Review

REVIEW QUESTIONS

I. Matching

A. Nucleus   C. Golgi apparatus   E. Mitochondrion
B. Lysosome   D. Endoplasmic reticulum

B 1. This organelle is basically a package of digestive enzymes which are used intracellularly. *B*

E 2. The principal function of this organelle is energy transformation and the manufacture of high-energy phosphates *E*

C 3. A protein hormone made by the cell is packaged in final form for secretion by this organelle. *C*

D 4. It is associated with ribosomes and protein synthesis. Most proteins made by the cell are put together in this organelle. *D*

A 5. This organelle contains the genetic blueprints for protein synthesis. *A*

II. Multiple choice

A 6. Following are five steps in the scientific method. Which step is normally the second step?
A. hypothesis
B. observation
C. experiment
D. publication
E. conclusion

D 7. Which of the following is (are) considered (a) cell inclusion(s)?
A. nucleus
B. smooth ER
C. rough ER
D. pigment granule
E. lysosome

D 8. The term "pseudopodium" is used with reference to:
A. cell secretion
B. production of cellular ATP
C. ciliary movement
D. ameboid movement
E. the stadium of the Colts

___A___ 9. In a negative feedback control system:
A. the variable is returned toward normal after a disturbance
B. the system operates to destabilize
C. the variable never changes
D. when the value of the variable increases the system operates to increase it further
E. none of the preceding are true

___C___ 10. Positive feedback control systems:
A. add stability to the body's internal environment
B. tend to restore the normal value of an abnormal controlled variable
C. are not common in the body because they act to destabilize or reinforce the direction a variable is moving
D. involve only the brain and processes of learning
E. raise blood sugar level when it falls below normal

___A___ 11. B The smallest group of different types of cells that share common functional objectives as regards homeostasis are called a(an):
A. tissue   D. organelle
B. organ   E. membrane
C. system

___C___ 12. Which of the following is an inclusion?
A. lysosome   D. Golgi apparatus
B. endoplasmic reticulum   E. nucleus
C. pigment granule

___C___ 13. Energy transformation and ATP synthesis primarily occurs in this organelle:
A. lysosome   D. nucleus
B. golgi apparatus   E. centrosome
C. mitochondrion

___D___ 14. The principal function of the rough endoplasmic reticulum is to:
A. package carbohydrates prior to secretion
B. generate ATP
C. duplicate chromosomes
D. manufacture proteins
E. digest macromolecules

outside protein
___C___ 15. Ectoproteins:
A. are proteins bound to the inner (intracellular) surface of the plasma membrane
B. may form a channel through which solute can diffuse into or out of the cell
C. may have a sugar attached (glycoprotein) that allows the molecule to act as a receptor for incoming signals
D. are the same as integral proteins
E. serve to bind together the two halves of the lipid bilayer

15

B 16. According to its literal definition, the word "physiology" means the study of:
A. animals
B. nature
C. plants
D. function
E. structure

D 17. "To remain the same" is a literal definition of the word:
A. homogenized
B. homosexual
C. homeopathic
D. homeostasis
E. homo sapiens

C 18. The organelle where most of the cell's proteins are synthesized is the:
A. nucleus
B. lysosome
C. rough ER
D. smooth ER
E. centrosome

A 19. In or on the phospholipid bilayer of a plasma membrane:
A. phospholipids are distributed asymmetrically
B. proteins are distributed asymmetrically
C. both proteins and phospholipids move laterally
D. glycoproteins are located on the extracellular surface
E. all of the above are true

E 20. A spherical membranous organelle containing digestive enzymes which are used by the cell to break down (catabolize) large molecules. The organelle is:
A. the smooth ER
B. a microtubule
C. a mitochondrion
D. the nucleus
E. a lysosome

C 21. This molecule is part carbohydrate and part protein. It is located on (in) the outer (extracellular) half of the plasma membrane. The molecule is a:
A. phospholipid
B. lipoprotein
C. glycoprotein
D. glycolipid
E. steroid

B 22. Which of the following is a cell inclusion?
A. lysosome
B. pigment granule
C. ribosome
D. endoplasmic reticulum
E. microtubule

D 23. "Pseudopodium" is most closely associated with:
A. ciliary movement
B. protein synthesis
C. carbohydrate synthesis
D. ameboid movement
E. bacterial infection

_E_ 24. Which of the following statements is true?
A. All mature human cells are mononucleate.
B. All mature human cells are multinucleate.
C. All mature human cells are anucleate (i.e., without a nucleus).
D. All mature human cells have at least one nucleus.
E. Most mature human cells are mononucleate, however some are multinucleate and anucleate.

_D_ 25. Cells that are very active metabolically and perform chemical and mechanical work usually have numerous:
A. centrioles
B. golgi apparati
C. lysosomes
D. mitochondria
E. cilia

_C_ 26. Proteins such as hormones are synthesized within a cell by the:
A. nucleus
B. Golgi apparatus
C. rough endoplasmic reticulum
D. lysosome
E. plasma membrane

_C_ 27. Which of the following best fits the modern definition of physiology?
A. the study of nature
B. the study of body parts
C. the study of body functions
D. the study of body chemistry
E. the study of animals

_D_ 28. The phase of mitosis in which the parent cell divides into two daughter cells is called:
A. metaphase
B. anaphase   cytoKenisis
C. prophase
D. telophase
E. interphase

_F_ 29. Duplication of chromosomes occurs during:
A. anaphase
B. prophase
C. metaphase
E. telophase
F. interphase

_B_ 30. Secretion of a hormone by an endocrine cell is usually a process of:
A. net diffusion   D. phagocytosis
B. exocytosis   E. pinocytosis
C. endocytosis

OUTLINE 2

LEARNING OBJECTIVES(√)

After reading the assigned pages in the textbook, and/or reading and performing related laboratory exercises, and listening to lecture and/or laboratory presentations, you should be able to:

☐1. Explain the difference between a concentration difference and a concentration gradient.

☐2. Compute the value of a concentration gradient if given values for concentrations and the length of the diffusion pathway.

☐3. Describe the relation between the rate of net diffusion of a substance and each of the factors in Fick's law.

☐4. Define "directly proportional" and "inversely proportional" as used to describe the influence of factors in Fick's law on the rate of net diffusion of a solute.

☐5. Define the term "permeability" and the term "lipid solubility".

☐6. Explain how each of the following influences net diffusion through a plasma membrane: lipid solubility, molecular size and shape, and electrical charge.

☐7. Define osmosis and dialysis in terms of net diffusion.

☐8. Explain why osmosis requires the presence of a non-diffusible solute on one side of a selectively permeable membrane but dialysis does not require the non-diffusible solute.

☐9. Explain how a hydrostatic pressure can be used to stop osmosis.

☐10. Define the terms hypertonic, isotonic, and hypotonic, and explain what is meant by "physiological saline".

☐11. Explain what will happen to red blood cells if a hypertonic solution is infused into the blood stream of a normal person.

☐12. Define crenation and hemolysis.

☐13. Explain the relation of non-diffusible solute concentration to the osmotic pressure of the solution.

☐14. List and explain four general characteristics of plasma membrane carrier-mediated transport systems.

☐15. Describe the differences between carrier-mediated facilitative diffusion and primary active transport.

☐16. Define symport and antiport and give an example of each.

☐17. Explain the difference between phagocytosis and pinocytosis.

☐18. Explain the difference between endocytosis and exocytosis.

NOTES

OUTLINE #_____

QUESTIONS

1._____
2._____
3._____
4._____
5._____

OUTLINE 2

I. **NET DIFFUSION**

    A. Gradient = Difference/Distance

    B. Fick's Law: Concentration Difference, Cross-sectional Area, Temp., Distance, Molecular Weight

        Equation:

    C. Diffusion Through Plasma Membrane

        1. Permeability

        2. Lipid solubility

        3. Solute size and shape

        4. Electrical charge

    D. Osmosis

        1. Definition

        2. Osmotic pressure

            a. Isotonic : ICF osmotic pressure = ECF osmotic pressure

            b. Hypertonic: ICF osmotic pressure > ECF osmotic pressure

            c. Hypotonic: ICF osmotic pressure< ECF osmotic pressure

    E. Dialysis

        1. Definition

        2. Hemodialysis
           Diagram: elements of a dialyzing device

Hemolysis

NOTES

DATE : _____   *OUTLINE #*_____

QUESTIONS

1._____
2._____
3._____
4._____
5._____

II. CARRIER-MEDIATED TRANSPORT

A. General Characteristics of Carriers
   1. Chemistry

   2. Specificity

   3. Saturation

   4. Competition

B. Facilitated Diffusion (carrier-mediated)
   Diagram: carrier moving solute along gradient

C. Active Transport
   Diagram: carrier moving solute against gradient

III. ENDOCYTOSIS
A. Phagocytosis *Eat*

B. Pinocytosis *Drink*
   Diagram: leukocyte phagocytosis

IV. EXOCYTOSIS
Diagram: Golgi apparatus and cell secretion
A. Secretion

B. Excretion

NOTES

DATE : _____   *OUTLINE #*_____

QUESTIONS

1._____
2._____
3._____
4._____
5._____

Difference vs. Gradient

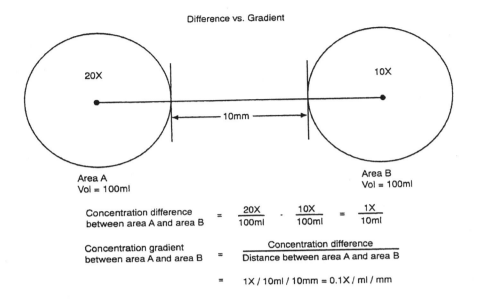

Area A
Vol = 100ml

Area B
Vol = 100ml

Concentration difference between area A and area B $= \dfrac{20X}{100ml} - \dfrac{10X}{100ml} = \dfrac{1X}{10ml}$

Concentration gradient between area A and area B $= \dfrac{Concentration\ difference}{Distance\ between\ area\ A\ and\ area\ B}$

$= 1X / 10ml / 10mm = 0.1X / ml / mm$

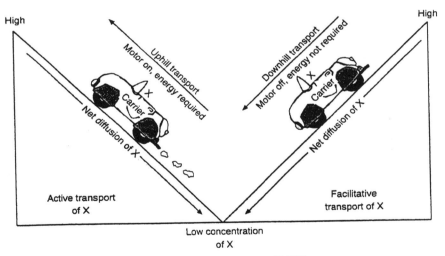

CARRIER-MEDIATED TRANSPORT

NOTES

OUTLINE #_____

QUESTIONS

1._____
2._____
3._____
4._____
5._____

REVIEW QUESTIONS

I.   Matching

    A.   Energy-dependent carrier – mediated transport

    B.   Net diffusion

    C.   Facilitative diffusion

    D.   Osmosis

    E.   Dialysis

A 1.   Requires a membrane; is directly coupled to an energy-yielding reaction; moves solute against gradient. *A*

B 2.   Solute movement from higher to lower solute concentration without membrane. *B*

D 3.   Solvent movement across membrane from higher to lower solvent concentration. *D*

C 4.   Solute movement by carrier across membrane from higher to lower solute concentration. *C*

E 5.   Net diffusion of solute across a selectively permeable membrane. *E*

II.   Multiple choice

C 6.   Which of the following processes requires a supply of metabolic energy?

    A.   simple diffusion

    B.   facilitated diffusion

    C.   primary active transport

    D.   dialysis

    E.   osmosis

E 7.   Facilitated diffusion differs from simple diffusion because:

    A.   it is much more rapid

    B.   there is a limit to its rate

    C.   movement can be blocked by competitive inhibitors

    D.   it is highly specific

    E.   all of the above

D 8.   The net diffusion of solute through a selectively permeable membrane describes:

    A.   osmosis

    B.   active transport

    C.   facilitated diffusion

    D.   dialysis

    E.   symport

_D_ 9.   Placing a RBC into this solution will result in the cell crenating. This solution is:
A.   .9% NaCl
B.   .2% NaCl
C.   hypotonic
D.   hypertonic
E.   two of the above

_C_ 10.   One factor that influences the net diffusion of a solute but is not included in Fick's law is:
A.   the concentration difference of the solute
B.   the cross-sectional area of the diffusion pathway
C.   the lipid solubility of the solute
D.   the length of the diffusion pathway
E.   the molecular weight of the solute

_D_ 11.   Regarding osmosis and dialysis, which of the following is false?
A.   both require a membrane
B.   both involve net diffusion
C.   neither require a membrane carrier
D.   both involve solute movement
E.   neither require the energy from ATP

_B_ 12.   Primary active transport systems:
A.   are always coupled to Na⁺ movement
B.   are always directly coupled to energy-yielding reactions
C.   require energy but are not usually coupled to energy-yielding reactions
D.   always carry K⁺ into the cell
E.   can best be described by the term "symport"

_C_ 13.   Examination of human red blood cells, after removing them from the body and placing them in solution, reveals widespread crenation. With respect to the plasma, the solution that the cells were placed in must have been:
A.   isotonic
B.   hypotonic
C.   hypertonic
D.   hypoosmotic
E.   two of the preceding

_C_ 14.   One factor that influences the rate of net diffusion of a solute but is not included in Fick's law is:
A.   the concentration difference of the solute
B.   the cross-sectional area of the diffusion pathway
C.   the electrical charge of the diffusing solute
D.   the length of the diffusion pathway
E.   the molecular weight of the diffusing solute

notes

___D___ 15. Molecules of X net diffuse from area A to area B, and molecules of Y net diffuse from area B to area A in the same system. The molecular weight of X is 696.68, and the molecular weight of Y is 319.86. According to Fick's law of diffusion:
A. X will net diffuse faster than Y
B. X is more soluble than Y
C. assuming the concentration gradients for X and Y are equal, X will achieve diffusion equilibrium faster than Y
D. an increase in system temperature will increase the net diffusion of both X and Y and result in diffusion equilibrium occurring in a shorter period of time
E. none of the preceding statements are true

___C___ 16. The concentration of A in area X is 60 mg/ml. The concentration of A in area Y is 20 mg/ml. The length of the diffusion pathway that connects area X to area Y is 20 millimeters. The concentration gradient for the net diffusion of A from X to Y is:
A. 20 mg/mm
B. 10 mg/ml/mm
C. 2.0 mg/ml/mm
D. 1mg/ml/mm
E. 1ml/mm

___E___ 17. According to Fick's Law:
A. Anything that can go wrong will.
B. What goes up must come down.
C. If it isn't broke, don't fix it.
D. Substances that are lipid soluble enter cells more easily than substances that aren't lipid soluble.
E. The rate of net diffusion of a solute from higher to lower concentration is directly proportional to the magnitude of the concentration difference.

___A___ 18. A membrane which is permeable only to water separates two solutions of glucose dissolved in water. On one side of the membrane (side A) the glucose concentration is 0.1 g/ml. On the other side (side B) the glucose concentration is 0.5 g/ml. Initially the rate of water flow will be:
A. most rapid from side A to side B
B. most rapid from side B to side A
C. the same in both directions
D. zero (no flow in either direction)
E. from higher to lower solute concentration

___B___ 19. Which of the following statements is true?
A. Secondary active transport pumps are directly connected to energy-yielding reactions.
B. Facilitative diffusion involves proteins.
C. Osmosis requires net diffusion of a solute.
D. Primary active transport systems are not energy dependent.
E. All of the above are true.

_D_ 20.   Dialysis:
- A.   does not require a membrane
- B.   involves "up-hill" transport of solute
- C.   requires membrane carriers
- D.   requires a selectively permeable membrane
- E.   two of the preceding

_C_ 21.   Plasma membrane transport molecules (carriers):
- A.   are usually lipids
- B.   generally transport many types of solutes (i.e. the carriers are not very specific)
- C.   are proteins that remain confined to the plasma membrane
- D.   transport water as well as solute
- E.   are not involved in facilitated diffusion

_C_ 22.   Antiport refers to carrier transport of a solute:
- A.   down an electrical gradient
- B.   coupled to and in the same direction as Na^+ transport
- C.   coupled to but in the opposite direction as Na^+ transport
- D.   against an osmotic gradient
- E.   through the nuclear envelope

_C_ 23.   Invagination of the plasma membrane, after a surface receptor has been engaged, and formation of an endocytotic vesicle (phagosome) is an example of:
- A.   exocytosis
- B.   secretion
- C.   phagocytosis
- D.   pinocytosis
- E.   amitosis

_E_ 24.   Which of the following statements is false?
- A.   Osmosis can cause cells to swell.
- B.   Osmosis involves movement of water through specific water channels.
- C.   Osmosis may involve large scale movement of water molecules.
- D.   Osmosis is involved in meat preservation by salting.
- E.   Osmosis is water flow from high solute to low solute concentration.

_A_ 25.   Simple (net) diffusion differs from facilitated diffusion because it:
- A.   is much slower
- B.   can be blocked by competitive inhibitors
- C.   reaches a maximum rate
- D.   requires expenditure of energy
- E.   moves solute against a gradient

C 26. Which of the following processes requires a direct supply of metabolic energy?
- A. simple diffusion
- B. facilitated diffusion
- C. primary active transport
- D. osmosis
- E. diffusion trapping

C 27. Symport refers to:
- A. carrier transport of molecule X in the opposite direction molecule Y is being transported
- B. two different carriers transporting molecules of X in the same direction across the plasma membrane
- C. a single carrier transporting molecule X and molecule Y in the same direction across the plasma membrane
- D. two different carriers transporting molecules of X in one direction and molecules of Y in the opposite direction
- E. none of the above

D 28. Which of the following statements is false?
- A. Osmosis requires a selectively permeable membrane.
- B. Dialysis requires a selectively permeable membrane.
- C. Carrier-mediated transport requires a biological membrane.
- D. Net diffusion requires a biological membrane but not an artificial membrane.
- E. Facilitative diffusion requires a membrane

D 29. The type of molecule most likely to rapidly cross a plasma membrane by diffusing through a protein pore is a molecule that is:
- A. large and positively charged
- B. small and negatively charged, like chloride ions
- C. soluble in lipids
- D. small and polar, like water
- E. large and non-polar

B 30. A red blood cell is placed in a 0.9% sodium chloride solution and intracellular volume does not change because:
- A. the salt solution is hypotonic
- B. the salt concentration inside the cell is 0.9%
- C. the salt solution is hypertonic
- D. the plasma membrane is impermeable to water
- E. intracellular and extracellular fluids are iso-osmotic

NOTES

DATE : _____   OUTLINE #_____

QUESTIONS

1._____
2._____
3._____
4._____
5._____

OUTLINE 3

LEARNING OBJECTIVES (√)

After reading the assigned pages in the textbook, and/or reading and performing related laboratory exercises, and listening to lecture and/or laboratory presentations, you should be able to:

☐1. Define resting membrane potential and give examples of its variability.

☐2. Use a diagram of a generalized animal cell to explain the genesis of the resting membrane potential.

☐3. Define equilibrium potential and explain what is meant by electrochemical equilibrium as regards the chemical and electrical diffusion gradients through a plasma membrane for a diffusible ion such as potassium.

☐4. Explain the role of the Na^+/K^+ pump in the maintenance of the resting membrane potential, and how the potential would change if the pumps stopped.

☐5. Define depolarization and hyperpolarization as they apply to changes in the plasma membrane resting potential.

☐6. Define neuron and neuroglia.

☐7. Diagram a generalized neuron including the following elements: dendrites, soma, axon, myelin sheath, node of Ranvier, terminal boutons, and a synapse.

☐8. Diagram three types of neurons (unipolar, bipolar, and multipolar) and indicate dendrite, soma, and axon for each type.

☐9. Diagram an action potential and explain the changes in ion conductance that occur at stimulus application, threshold potential, rapid depolarization, repolarization, and hyperpolarization.

☐10. Describe propagation of the action potential (conduction of the nerve impulse).

☐11. Define the "all or none" law as it applies to the amplitude and conduction of the action potential.

☐12. Explain absolute refractory period and relative refractory period.

☐13. List and explain the influence of two factors on the conduction velocity of a nerve impulse.

☐14. Explain three ways information regarding the type, intensity, and location of a somatic sensory stimulus can be coded.

NOTES

QUESTIONS

1._____
2._____
3._____
4._____
5._____

OUTLINE 3

LECTURER:_____   NOTES_____   DATE:_____

I. PLASMA MEMBRANE ELECTRICAL POTENTIALS

A. Resting Membrane Potential
   1. Definition

   2. Examples of variability

   3. Genesis of the resting membrane potential (E_r)
      Diagram: development of a resting membrane potential

   4. Electrochemical equilibrium: chem. gradient/elect. gradient

   5. Maintenance of E_r: Na^+/K^+ Pump

B. Change in the Resting Membrane Potential
   1. Depolarization: definition

   2. Hyperpolarization: definition   drops below norm.
      after polarization K^+?

II. NEURONS: STRUCTURAL AND FUNCTIONAL ORGANIZATION

Diagram: basic structure of a multipolar neuron
   A. Dendrites - recieves signals - extends from soma
   B. Soma - neuron cell body - metabolic center
   C. Axon - sends signal - ext. from soma
   D. Myelin - insulates axon to make transmission faster
   E. Node of Ranvier - space between mylination
   F. Terminal Bouton - End of axon where neurotransmitters
      are sent out
   G. Synapse - junction between nerve cells

   axon hillock where action potential is generated

34

NOTES

1 trillion cells
NEURON - cell of nervous system - same as any

CNS - study cervical, thoracic, lumbar
        brain + spinal
PNS - peripherial
study - anatomical diagrams

neural tube = CNS
neural crest = PNS

Neurons - transmit + receive (no cancer)
don't reproduce

neuroglia - nurish, aid transmition
clean, insulate, reproduce (can be cancer)

QUESTIONS

1._____
2._____
3._____
4._____
5._____

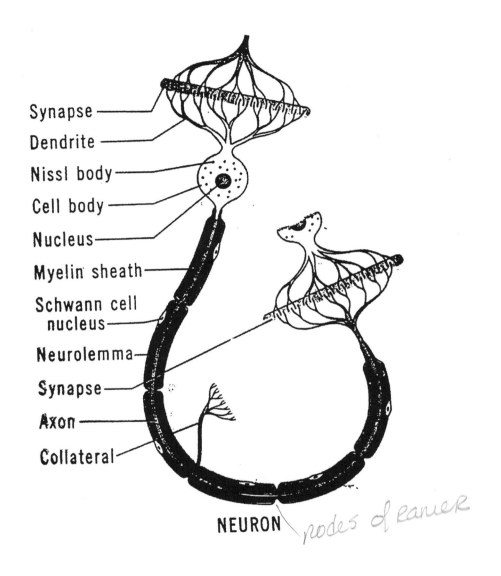

Synapse

Dendrite

Nissl body

Cell body

Nucleus

Myelin sheath

Schwann cell
nucleus

Neurolemma

Synapse

Axon

Collateral

NEURON *nodes of ranier*

NOTES

DATE : _____   *OUTLINE #_____*

QUESTIONS

1._____
2._____
3._____
4._____
5._____

III. **TYPES OF NEURONS**
   Diagram: *Unipolar*   *Bipolar*   *Multipolar*

IV. **CONDUCTION OF THE NERVE IMPULSE**
   A.   The Action Potential
       Diagram: generation of an action potential
       1.   Resting membrane potential of the neuron

       2.   Stimulus application and permeability changes
       3.   Threshold potential and depolarization
       4.   Spike potential and repolarization

   B.   Propagation of the Action Potential
       Diagram: propagation of nerve impulse; non-myelinated fiber
       1.   Local current flow
       2.   Depolarization wave/repolarization wave

       3.   All or none law

       4.   Refractory period: absolute and relative
           Diagram: appended

NOTES

QUESTIONS

1._____
2._____
3._____
4._____
5._____

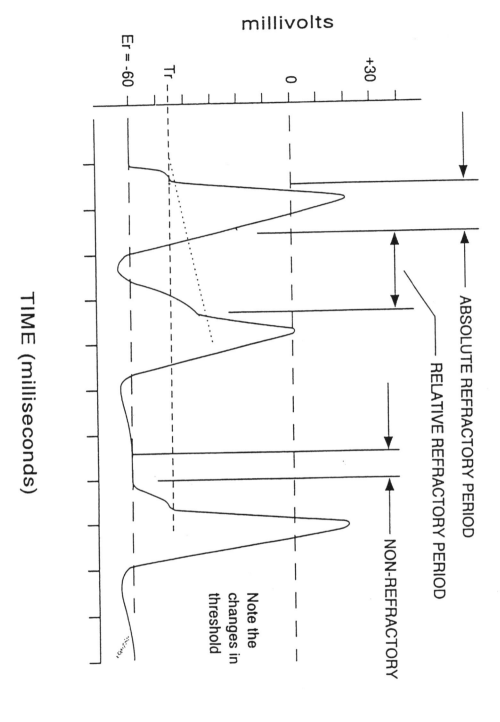

ACTION POTENTIALS

millivolts

Er = -60
Tr
0
+30

TIME (milliseconds)

ABSOLUTE REFRACTORY PERIOD

RELATIVE REFRACTORY PERIOD

NON-REFRACTORY

Note the
changes in
threshold

NOTES

DATE : _____   *OUTLINE #_____*

QUESTIONS

1._____
2._____
3._____
4._____
5._____

C. Conduction Factors
Diagram: propagation of nerve impulse by a myelinated fiber
    1.   Myelin and saltatory conduction

    2.   Fiber diameter

D. Information Coding
    1.   Frequency code (# of impulses per second)

    2.   Population code (varying # of simultaneously active nerve fibers)

    3.   Labeled line code (specific pathways carry specific information)

NOTES

DATE : _____   *OUTLINE #_____*

QUESTIONS

1._____
2._____
3._____
4._____
5._____

REVIEW QUESTIONS

I. Matching

    A. Dendrite
    B. Soma
    C. Axis cylinder (axon)
    D. Terminal bouton
    E. Two of the above

___1. short branch-like processes where incoming signals are received

___2. tiny button-like swelling where presynaptic release of neurotransmitter occurs

___3. where the action potential first appears

___4. the part of the neuron where integration of excitatory and inhibitory signals occurs.

___5. conduction of the action potential away from the neuron cell body

II. Multiple choice

___6. The time interval during which no stimulus can elicit an action potential in a nerve fiber is called the:
    A. latent period
    B. relative refractory period
    C. absolute refractory period
    D. repolarization period
    E. threshold

___7. In the sodium-potassium pump:
    A. sodium and potassium both leave the cell
    B. sodium and potassium both enter the cell
    C. sodium enters and potassium leaves the cell
    D. potassium enters and sodium leaves the cell
    E. three potassium ions are exchanged for one sodium ion

___8. The principal cause of early repolarization of a nerve fiber after an adequate stimulus has been applied is:
    A. an increase in the inward conductance of K^+
    B. an increase in the outward conductance of Na^+
    C. an increase in the inward conductance of Na^+
    D. an increase in the outward conductance of K^+
    E. a decrease in the inward conductance of Na^+

___9. The principal cause of depolarization of a nerve fiber when an adequate stimulus is applied is:
    A. an increase in the inward conductance of K^+
    B. an increase in the outward conductance of Na^+
    C. an increase in the inward conductance of Na^+
    D. an increase in the outward conductance of K^+
    E. a decrease in the inward conductance of Na^+

_B_ 10. Saltatory conduction refers to:
- A. conduction between presynaptic and postsynaptic membranes
- B. conduction of a nerve impulse by a myelinated axon
- C. conduction of a signal between nerve and muscle
- D. conduction or electrical current through body water
- E. conduction of an impulse by muscle T-tubules

_B_ 11. The period of excitability during which a greater than normal strength of stimulus is required to elicit a cell response is termed the:
- A. interphase period
- B. relative refractory period
- C. absolute refractory period
- D. hyperexcitable period
- E. none of the preceding

_E_ 12. The establishment and maintenance of a cell's resting membrane potential is dependent on:
- A. selective permeability of the membrane to ions
- B. active transport of ions
- C. the presence of non-diffusible anions inside the cell
- D. energy expenditure
- E. all of the above

_B_ 13. The hyperpolarization phase of the action potential is caused by too much _____ leaving the cell
- A. sodium ion
- B. potassium ion
- C. calcium ion
- D. phosphate ion
- E. water

_D_ 14. According to the "all or none" law:
- A. the amplitude of the action potential is directly proportional to the strength of the applied stimulus
- B. the amplitude of the action potential is inversely proportional to the strength of the applied stimulus
- C. the speed of nerve impulse conduction is inversely proportional to the diameter of the nerve fiber
- D. for a given nerve fiber, the amplitude of the action potential does not vary with the strength of the stimulus
- E partial depolarization of the neuron's plasma membrane never occurs

_D_ 15. The resting membrane potential of a neuron is dependent on:
- A. active transport of potassium ions out of the cell
- B. active transport of sodium ions into the cell
- C. the opening of electrically operated sodium channel gates
- D. the presence of non-diffusible anions inside the cell
- E. none of the above.

_D_ 16. According to the "all or none" law:
  A. saltatory conduction is faster
  B. the stronger the stimulus the greater the amplitude of the action potential
  C. either all of the fibers in a nerve are active at the same time or none of them are
  D. action potential amplitude for a given neuron is constant
  E. none of the above are true

_D_ 17. With respect to the neuron, the sodium - potassium pump:
  A. pumps sodium in, potassium out
  B. generates ATP
  C. keeps the resting membrane potential near zero
  D. maintains the outward potassium diffusion gradient
  E. becomes inoperative (i.e., stops) during repolarization

_C_ 18. In which of the following would the velocity of nerve impulse conduction be the greatest?
  A. large diameter unmyelinated fibers
  B. small diameter unmyelinated fibers
  C. large diameter myelinated fibers
  D. small diameter myelinated fibers
  E. none of the preceding; they conduct impulses at the same velocity

_E_ 19. What is the direction of the driving force(s) for the movement of potassium ions when a nerve cell is at rest?
  A. inward electrical gradient
  B. inward chemical gradient
  C. outward electrical gradient
  D. outward chemical gradient
  E. both A and D

_B_ 20. If energy input (in the form of ATP) to the Na^+ - K^+ pump ceases:
  A. the resting membrane potential becomes larger
  B. the resting membrane potential goes to zero
  C. an EPSP develops
  D. the cell loses sodium
  E. two of the preceding

_B_ 21. Information is coded in the nervous system by:
  A. altering the amplitude of the action potential
  B. varying the frequency of the action potentials
  C. increasing the velocity of impulse conduction
  D. changing the type of neurotransmitter used by a neuron
  E. two of the preceding

_B_ 22. The most common type of neuron in the central nervous system (brain and spinal cord) is the _____ neuron.
  A. unipolar   C. bipolar   E. dendritic
  B. multipolar   D. nonpolar

23. Sensory information can be coded by varying the number of receptors that fire in response to a stimulus of a given intensity. This type of code is called a:
    A.   frequency code
    B.   population code
    C.   labeled line code
    D.   Morse code
    E.   Alphanumeric code

24. Terminal boutons are part of the _____ of the neuron.
    A.   dendrite
    B.   soma
    C.   axon
    D.   myelin sheath
    E.   post-synaptic membrane

25. The normal action potential exhibited by a single neuron:
    A.   has a variable threshold potential
    B.   has a variable amplitude
    C.   is conducted without decrement (decreased velocity/amplitude)
    D.   may or may not be propagated (conducted)
    E.   is generated on the terminal bouton

26. What is the direction of the driving forces for the movement of sodium ions at the peak of the action potential spike?
    A.   inward electrical gradient
    B.   outward electrical gradient
    C.   inward chemical gradient
    D.   outward chemical gradient
    E.   both (B) and (C)

27. If the Na^+/K^+ pumps in a nerve fiber plasma membrane were stopped, the electrical potential of the membrane would:
    A.   increase
    B.   decrease toward zero but remain negative inside, positive outside
    C.   decrease toward zero but remain positive inside, negative outside
    D.   become zero
    E.   be unaffected

28. The hyperpolarization phase of the action potential is caused by too much _____ leaving the cell.
    A.   sodium ion
    B.   potassium ion
    C.   calcium ion
    D.   phosphate ion
    E.   water

OUTLINE 4

LEARNING OBJECTIVES (√)

After reading the assigned pages in the textbook, and/or reading and performing related laboratory exercises, and listening to lecture and/or laboratory presentations, you should be able to:

☐1. Define ephaptic synapse and explain how an electrochemical impulse is passed from presynaptic cell to postsynaptic cell.

☐2. Diagram an electrochemical synapse including the following components: pre and postsynaptic membranes, synaptic cleft, storage vesicles of neurotransmitter, postsynaptic membrane receptors for neurotransmitter molecules.

☐3. Explain, with the aid of a diagram, how depolarization of the presynaptic membrane alters Ca^{++} flux and initiates vesicle movement in the presynaptic terminal.

☐4. Explain the relation between the frequency of presynaptic membrane depolarization and the amount of neurotransmitter released during synaptic transmission.

☐5. Define Excitatory Postsynaptic Potential (EPSP) and explain the change(s) in ion conductance that occur when the postsynaptic receptors are engaged by neurotransmitter.

☐6. Define Inhibitory Postsynaptic Potential (IPSP) and explain the change(s) in ion conductance that occur when the postsynaptic receptors are engaged by neurotransmitter.

☐7. Use acetylcholine as an example to explain why and how a neurotransmitter molecule is removed from a postsynaptic membrane receptor and recycled.

☐8. Explain why the axon hillock (initial segment) is important in the genesis of an action potential, i.e. why does the nerve impulse start here?

☐9. Define temporal summation and use a diagram to explain the process.

☐10. Define spatial summation and use a diagram to explain the process.

☐11. Explain the process of neuronal integration in which excitatory (+) and inhibitory (-) inputs occurring simultaneously are summated to determine postsynaptic cell firing (yes or no) and if yes, the frequency (the number of impulses per second) of firing.

NOTES

QUESTIONS

1._____
2._____
3._____
4._____
5._____

OUTLINE 4

LECTURER:_____   NOTES_____   DATE:_____

I. **SYNAPTIC TRANSMISSION**
    A. Ephatic vs. Electrochemical Synapse: Functional Difference

    B. Structure of the Ephaptic Synapse
        Diagram: current flow at ephaptic synapse

    C. Structure of the Electrochemical Synapse
        Diagram: elements of an electrochemical synapse
        1. Presynaptic membrane
        2. Synaptic cleft
        3. Postsynaptic membrane
        4. Storage vesicles of neurotransmitter
        5. Postsynaptic receptors

    D. Transmission Events
        1. Depolarization of terminal bouton and Ca^{++} flux

        2. Vesicle migration and exocytosis

        3. Engagement of postsynaptic receptors by transmitter and post-synaptic membrane permeability change
            a. Excitatory postsynaptic potential (EPSP)
                Diagram: ion movements: depolarization

            b. Inhibitory postsynaptic potential (IPSP)
                Diagram:: ion movements: hyperpolarization

NOTES

DATE : _____   *OUTLINE # _____*

QUESTIONS

1._____

2._____

3._____

4._____

5._____

4.   Degradation and recycling of neurotransmitter molecules
    a.   Enzymes and presynaptic uptake

    b.   Blocking synaptic transmission

II.   NEURON INTEGRATION

A.   Axon Hillock (Initial Segment) and Genesis of Action Potential
Diagram: current flow from postsynaptic membrane to initial segment

B.   Temporal Summation
Summation of two successive inputs at the same synapse
Diagram: somatic integration involving temporal summation

C.   Spatial Summation
Summation of two or more separate inputs at different synapses
Diagram: somatic integration involving spatial summation

NOTES

DATE : _____   *OUTLINE #_____*

QUESTIONS

1._____
2._____
3._____
4._____
5._____

SPATIAL SUMMATION

Spatial Has to be = Neg + Pos

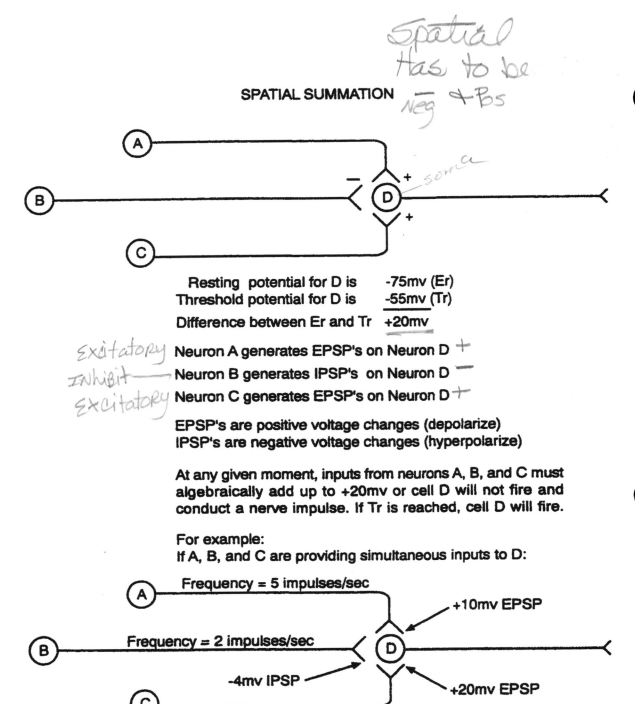

soma

Resting potential for D is   -75mv (Er)
Threshold potential for D is   -55mv (Tr)
Difference between Er and Tr   +20mv

Excitatory   Neuron A generates EPSP's on Neuron D +
Inhibit   Neuron B generates IPSP's on Neuron D —
Excitatory   Neuron C generates EPSP's on Neuron D +

EPSP's are positive voltage changes (depolarize)
IPSP's are negative voltage changes (hyperpolarize)

At any given moment, inputs from neurons A, B, and C must
algebraically add up to +20mv or cell D will not fire and
conduct a nerve impulse. If Tr is reached, cell D will fire.

For example:
If A, B, and C are providing simultaneous inputs to D:

	+10 mv
Ⓐ	+10 mv
Ⓒ	+ 20 mv
Ⓑ	— 4mv
+26mv net EPSP	

Therefore cell D will fire.
Spatial summation: the algebraic addition of inputs
from 2 different synapses on the same neuron

54

NOTES

DATE : _____   *OUTLINE #_____*

QUESTIONS

1._____
2._____
3._____
4._____
5._____

REVIEW QUESTIONS

I. Matching

A. Graded; non-propagated; increased chloride conductance into post-synaptic cell

B. Non-graded; propagated; all or none law

C. Non-propagated; graded; increased sodium conductance into post-synaptic cell which is not skeletal muscle

D. NM junction; increased sodium ion conductance into postsynaptic cell

E. Stable; maintained by the Na^+/K^+ pump

_D_ 1. End plate potential _D_
_A_ 2. Inhibitory postsynaptic potential _A_
_C_ 3. Excitatory postsynaptic potential _C_
_B_ 4. Action potential _B_
_E_ 5. Resting membrane potential _E_

II. Multiple choice

_E_ 6. Which of the following statements is true?
   A. For any given neuron, synaptic vesicles usually contain several kinds of neurotransmitters.
   B. For any given effector, the postsynaptic membrane has receptors for only one kind of neurotransmitter.
   C. Ephaptic synapses require neurotransmitter molecules.
   D. The amount of neurotransmitter released into the synaptic cleft is directly proportional to the amplitude of the nerve impulse.
   E. Statements A, B, C, and D are false.

_E_ 7. Synaptic transmission can be blocked by:
   A. blocking the transmitter synthesis
   B. blocking transmitter release
   C. blocking the postsynaptic receptor
   D. blocking the enzyme that degrades the neurotransmitter
   E. all of the above

_B_ 8. A neuron is receiving two simultaneous inputs, one of which is (-) 10mv. If the cell's E_r = (-)70mv and the T_r = (-) 60mv, what is the minimum value of the other input that would cause the neuron to generate an action potential?
   A. (+) 10mv   _-10_
   B. (+) 20mv
   C. (-) 10mv
   D. (-) 20mv
   E. (+) 30mv

56

_B_9. Which of the following is a graded, non-propagated, hyperpolarization of a neuron plasma membrane?
  A. EPSP
  B. IPSP
  C. EPP
  D. AP
  E. TP

_D_10. The most probable site on a neuron where incoming information is received is the:
  A. terminal bouton
  B. axon
  C. perikaryon
  D. dendrite
  E. myelin sheath

_D_11. Depolarization of a dendrite results in current flow between the dendrite and the _____ resulting in propagation of a nerve impulse.
  A. adjacent dendrites
  B. terminal boutons
  C. perikaryon
  D. axon hillock
  E. myelin sheath

_D_12. During transmission at an electrochemical synapse, neurotransmitter molecules are released from the presynaptic neuron by:
  A. facilitative diffusion
  B. active transport
  C. endocytosis
  D. exocytosis
  E. phagocytosis

_B_13. Which of the following enters the postsynaptic terminal and initiates an EPSP at an electrochemical synapse?
  A. Cl^-
  B. Na^+
  C. K^+
  D. Ca^{++}
  E. H^-

_D_14. Synaptic vesicles are induced to migrate toward and blend with the presynaptic membrane and then to release neurotransmitter into the synaptic cleft as a result of an increase in the concentration of free ____ in the presynaptic terminal.
  A. K^+
  B. Na^+
  C. Cl^-
  D. Ca^{++}
  E. Ach

D 15.   Which of the following statements is true?
  A.   For any given neuron, synaptic vesicles may contain more than   one type of neurotransmitter.
  B.   For any given effector, the postsynaptic membrane has receptors   for only one kind of neurotransmitter.
  C.   Ephaptic synapses require neurotransmitter molecules.
  D.   Neurotransmitter molecules are stored in vesicles located in the terminal bouton region of the presynaptic cell.
  E.   Most of the neurotransmitter molecules are manufactured in the terminal bouton region.

A 16.   Somatic integration involving the additive effect of 2 or more EPSP's produced in rapid succession at one synapse defines:
  A.   temporal summation
  B.   spatial summation
  C.   motor unit summation
  D.   mechanical summation
  E.   none of the above

E 17.   Excitatory postsynaptic potentials occur when a neuro-transmitter engages a postsynaptic membrane receptor resulting in the membrane:
  A.   becoming more permeable to Cl^- in
  B.   becoming more permeable to K^+ in
  C.   becoming less permeable to Na^+ in
  D.   becoming less permeable to K^+ in
  E.   becoming more permeable to Na^+ in

E 18.   A nerve fiber has a resting potential of –80mv and a threshold potential of –60mv. Which of the following is most likely to generate an action potential?
  A.   +15mv EPSP
  B.   -15mv IPSP
  C.   +5mv EPSP
  D.   -20mv IPSP
  E.   A + C (sum)

C 19.   In a neuron, where does an action potential first appear after the neuron has been adequately stimulated on a dendrite?
  A.   the dendrite
  B.   the soma
  C.   the axon initial segment
  D.   the middle of the axon
  E.   the terminal bouton

C 20.   When the effects of two EPSP's occurring at different synapses combine to generate an action potential, the process is called:
  A.   synaptic summation   D.   somatic summation
  B.   temporal summation   E.   none of the above
  C.   spatial summation

_D_ **21.** The process whereby a neuron releases transmitter molecules into the synaptic cleft is called:
A.   net diffusion
B.   osmosis
C.   endocytosis
D.   exocytosis
E.   phagocytosis

_B_ **22.** Which of the following is a graded, non-propagated, hyperpolarization of a neuron plasma membrane?
A.   EPSP
B.   IPSP
C.   EPP
D.   AP
E.   TP

_D_ **23.** The most probable site on a neuron where incoming information is received is the:
A.   terminal bouton
B.   axon
C.   perikaryon
D.   dendrite
E.   myelin sheath

_D_ **24.** Depolarization of a dendrite results in current flow between the dendrite and the _____ resulting in propagation of a nerve impulse.
A.   adjacent dendrites
B.   terminal boutons
C.   perikaryon
D.   axon hillock
E.   myelin sheath

_D A_ **25.** Integration of incoming signals is a function of which part of the neuron?
A.   dendrites
B.   terminal boutons
C.   axon
D.   perikaryon
E.   nucleus

_D_ **26.** Acetylcholinesterase
A.   is the neurotransmitter always found at the neuromuscular junction
B.   is a very common neurotransmitter found in central as well as peripheral nervous systems is an enzyme essential in the synthesis of acetylcholine
C.   is an enzyme essential in the synthesis of acetylcholine
D.   is an enzyme that cleaves acetylcholine from the receptor and splits the molecule into acetyl and choline portions
E.   has the same effect as curare (a neuromuscular blocker)

_C_ 27. Action potentials are normally generated at the beginning of the axon because:

A. myelin is absent

B. the membrane in this area lacks Na^+/K^+ pumps

C. the membrane in this area has a very low electrical resistance

D. there are very few Na^+ channels here

E. this part of the membrane is very permeable to K^+

_B_ 28. Which of the following governs the release of neurotransmitter from the presynaptic neuron?

A. the amplitude of the postsynaptic action potential

B. the frequency of the presynaptic action potentials

C. the number of postsynaptic engaged

D. Na^+ influx into the postsynaptic membrane

E. Cl^- efflux from the postsynaptic cell

_E_ 29. Which of the following temporal summations would be most effective in generating an action potential?

A. 5mV input, 1.0 sec delay, 10 mV input

B. 5mV input, 0.8 sec delay, 10 mV input

C. 5mV input, 0.6 sec delay, 10 mV input

D. 5mV input, 0.4 sec delay, 10 mV input

E. 5mV input, 0.1 sec delay, 10 mV input

_D_ 30. One of the following only represents spatial summation. The other choices could represent either spatial or temporal summation. Which one only represents spatial summation?

A. (+5mV) +(+10mV) Temporl

B. (+ 1mV) + (+ 2mV) Temporl

C. (- 5mV) + (-10mV) temporl

D. (-5mV) + (+10mV)

E. (-1mV) + (-2mV)

NOTES

QUESTIONS

1._____
2._____
3._____
4._____
5._____

61

OUTLINE 5

LEARNING OBJECTIVES (√)

After reading the assigned pages in the textbook, and/or reading and performing related laboratory exercises, and listening to lecture and/or laboratory presentations, you should be able to:

☐1. List three types of muscle tissue and their distinguishing features in terms of striation, general shape and size, location in the body, and innervation.

☐2. Outline the general gross structure of skeletal muscle, including fasciculi, fiber arrangements (parallel, pennate, bipennate, etc.), and tendons of origin and insertion.

☐3. Sketch and label the intracellular structure of a skeletal muscle fiber, including the following components: myofibril, sarcomere, actin myofilament, myosin myofilament, T-tubules, sarcoplasmic reticula, and Z discs.

☐4. Diagram the neuromuscular junction and use it to explain the steps of neuromuscular transmission.

☐5. Define end-plate potential and explain how it differs from a muscle fiber action potential.

☐6. Explain how neuromuscular transmission can be blocked by curare and why it is useful to be able to block neuromuscular transmission.

☐7. Explain how an action potential is generated on the sarcolemma in response to normal neuromuscular transmission.

☐8. Define excitation - contraction coupling.

☐9. Outline the roles of the transverse tubules (T-tubules), the sarcoplasmic reticula (SR), and Ca^{++} in excitation - contraction coupling.

☐10. Explain, in terms of function, the relation of troponin and tropomyosin to actin.

☐11. List, in the correct sequence, the events of one cross-bridge cycle, beginning with the charging of the myosin heads by ATP and the release of Ca^{++} from the sarcoplasmic reticula.

☐12. Define rigor mortis and explain, in relation to the cross-bridge cycle, why rigor mortis occurs after death.

NOTES

QUESTIONS

1._____
2._____
3._____
4._____
5._____

OUTLINE 5

LECTURER:_____   NOTES_____   DATE:_____

I.   **MUSCLE TISSUE**
    A.   General Function of Muscle:

    B.   General Comparison of Three Muscle Types
        Diagram: types of muscle cells
        1.   Smooth (nonstriated, involuntary, spindle-shaped, mononucleate)

        2.   Cardiac (striated, involuntary, cylindrical, mononucleate)

        3.   Skeletal (striated, voluntary, cylindrical, multinucleate

II. **SKELETAL MUSCLE**
    A.   Gross Anatomy
        1.   Fasciculus: bundles of skeletal muscle fibers

        2.   Fiber arrangements: parallel, pinnate, bipennate, multipennate, etc.

        3.   Tendons of origin and insertion

    B.   Cell Structure
        Diagram: organization of skeletal muscle cell
        1.   Fiber

        2.   Myofibril

        3.   Sarcomere

        4.   Myofilament

        5.   T-Tubules

        6.   Sarcoplasmic reticula

NOTES

QUESTIONS

1._____
2._____
3._____
4._____
5._____

C. Contraction and Relaxation

    1.   Neuromuscular junction transmission
        Diagram: motor endplate depolarization

    2.   T-Tubule depolarization and Ca^{++} release from SR
        Diagram: Ca^{++} release mechanism

    3.   The cross-bridge cycle:
        Diagram: skeletal muscle cross-bridge cycle
        (a)   Myosin head energized by ATP
        (b) Ca^{++} binds to troponin
        (c)   Troponin removes tropomyosin inhibition
        (d)   Myosin binds to actin, ATP $\rightarrow$ ADP + Pi + E released
        (e)   Myosin head rotates, pulling the attached actin filament
        (f)   ATP binds to myosin, myosin head detaches, ATP $\rightarrow$ ADP + Pi + E, head rotates back to resting position
        (g)   Cycle repeats as long as Ca^{++} is bound to troponin

NOTES

DATE : _____   OUTLINE #_____

QUESTIONS

1._____
2._____
3._____
4._____
5._____

REVIEW QUESTIONS

I. Matching

    A.   Sarcoplasmic reticulum
    B.   Troponin
    C.   Tropomyosin
    D.   ATP
    E.   Creatine phosphate

_A_ 1.   stores calcium ions *A*
_D_ 2.   the primary source of energy for cross-bridge cycling *D*
_C_ 3.   normally covers the actin active sites in the resting sarcomere *C*
_B_ 4.   binds calcium ion *B*
_E_ 5.   can be used to quickly generate ATP *E*

II. Multiple choice

_D_ 6. Ca^{++} is essential for skeletal muscle contraction because it bonds with and removes the inhibitory effect of:
    A.   actin
    B.   myosin
    C.   motor end plate
    D.   troponin-tropomysin complex
    E.   T- tubules

_B_ 7.   The function of troponin is to:
    A.   prevent actin and myosin from interacting
    B.   bind the calcium and release tropomyosin inhibition
    C.   store calcium ions until they are needed for contraction
    D.   act as a neurotransmitter between nerve and muscle
    E.   form cross-bridges with actin

_C_ 8.   The action potential generated near the neuromuscular junction is conducted into the skeletal muscle cell via:
    A.   actin filaments
    B.   sarcoplasmic reticula
    C.   T-tubules
    D.   myosin filaments
    E.   none of the preceding

_C_ 9.   Cross-bridge cycling in a skeletal muscle cell requires:
    A.   only ATP
    B.   only Ca^{++}
    C.   both ATP and Ca^{++}
    D.   acetylcholine
    E.   curare

_C_ 10. During the process of transmission at the neuromuscular junction, the role of acetylcholine is:
  A. to cause depolarization of the presynaptic (nerve) membrane
  B. to prevent depolarization of the muscle cell membrane
  C. to cause depolarization of the postsynaptic (endplate) membrane of the muscle cell
  D. block the effects of curare
  E. block the uptake of transmitter by the presynaptic cell

_B_ 11. The primary role of calcium in the activation of skeletal muscle is:
  A. to cause depolarization of the muscle cell plasma membrane
  B. to remove the inhibition of the reaction between the actin filaments and the myosin filaments
  C. to activate myosin molecules so that they can interact with actin filaments
  D. to conduct electrical impulses down the t-tubules
  E. release energy from mitochondria

_A_ 12. The role of the transverse tubules (T-tubules) in the excitation of skeletal muscle contraction is:
  A. to provide an inward path for the spread of muscle action potentials
  B. to serve as the storage site for calcium ions
  C. to connect the sarcomeres end to end
  D. to keep the muscle fiber from becoming fat in the middle and thin at the ends
  E. to maintain osmotic stability of the skeletal muscle fiber

_A_ 13. During excitation-contraction coupling in skeletal muscle:
  A. Ca^{++} is released from the SR
  B. Ca^{++} is pumped into the T-tubule
  C. ATP synthesis is depressed
  D. actin and myosin repel one another
  E. two of the preceding

_C_ 14. Sarcomere contraction involves:
  A. myosin filaments being pulled over actin filaments
  B. Ca^{++} binding to tropomyosin
  C. myosin heads attaching to actin and rotating about 45 degrees
  D. proteins of the Z bands moving apart
  E. two of the preceding

_D_ 15. During excitation-contraction coupling in skeletal muscle:
  A. the sarcoplasmic reticulum takes up Ca^{++}
  B. CA^{++} is pumped into the T-tubule
  C. $ATP + P + E \Rightarrow ADP$
  D. Ca^{++} is released from the sarcoplasmic reticulum
  E. the actin filaments uncoil

OUTLINE 6

After reading the assigned pages in the textbook, and/or reading and performing related laboratory exercises, and listening to lecture and/or laboratory presentations, you should be able to:

☐1. Define and diagram an example of a skeletal muscle motor unit.

☐2. Explain the relation between the size of each motor unit, the total number of motor units in a skeletal muscle, and the degree of neural control of the muscle.

☐3. Define motor unit recruitment and explain why and how motor unit recruitment is used by the brain to increase the strength of skeletal muscle contraction.

☐4. Diagram the record of a skeletal muscle twitch and label the contraction phase and the relaxation phase.

☐5. Explain the difference between isometric contraction and isotonic contraction of skeletal muscle.

☐6. Use diagrams of a sarcomere and a graph of force of contraction vs. initial length to explain how the initial length of a muscle is related to the muscle's ability to perform mechanical work.

☐7. Explain why energy contained in nutrient molecules such as glucose, amino acids, fatty acids or glycerol cannot be directly used to contract skeletal muscle.

☐8. Define oxygen debt and explain its origin in exercising skeletal muscle.

☐9. Use a sequence of chemical reactions to describe the relation of creatine phosphate to ATP and explain how creatine phosphate helps delay the onset of fatigue in a skeletal muscle.

☐10. List three physiological or biochemical differences between red slow twitch fibers, red fast twitch fibers, and white fast twitch fibers.

☐11. List two major causes of fatigue in skeletal muscle and explain why physical training increases the time to the onset of fatigue.

NOTES

DATE : _____   *OUTLINE #_____*

QUESTIONS

1._____
2._____
3._____
4._____
5._____

OUTLINE 6

LECTURER:_____ **NOTES**_____ **DATE:**_____

I. **MOTOR UNITS (MU)**
   A. Definition of Motor Unit:
   Diagram: motor unit = nerve fiber + muscle fibers

   B. Small MU and Large MU vs. Control of Skeletal Muscle

   C. Motor Unit Recruitment and Strength of Contraction

II. **SKELETAL MUSCLE TWITCH**
   A. Isometric vs. Isotonic Contraction
      1. Isometric contraction: iso *(equal)*, metric *(length)*

      2. Isotonic contraction: iso *(equal)*, tonic *(tension)*

   B. Initial Length vs. Strength of Contraction: Optimum Length
   Graph: Optimum length = maximum ability to do mechanical work

NOTES

DATE : _____   *OUTLINE #_____*

QUESTIONS

1._____
2._____
3._____
4._____
5._____

C.	Energy Sources
	1.	Adenosine triphosphate(ATP)
		Diagram: Formation of ATP from food metabolism

	2.	Creatine phosphate
		Diagram: Formation of ATP from CP metabolism

D.	Skeletal Muscle Fiber Types
	1.	Red slow twitch
		characteristics:

	2.	Red fast twitch
		characteristics:

	3.	White fast twitch
		characteristics:

## III.	EFFECT OF TRAINING ON SKELETAL MUSCLE
A.	Hypertrophy: Increase In Muscle Fiber Size

B.	Delayed Onset of Fatigue
	1.	Increased vascularity of trained skeletal muscle

	2.	Improved neuromuscular coordination

NOTES

DATE : _____   *OUTLINE #_____*

QUESTIONS

1._____
2._____
3._____
4._____
5._____

REVIEW QUESTIONS

I.   Matching

A.   Isotonic contraction
B.   Isometric contraction
C.   Isotonic relaxation
D.   Isometric relaxation
E.   Mixed contraction

_E_ 1.   most routine skeletal muscle contractions are of this type _F_
_B_ 2.   the first phase of a normal mixed contraction _B_
_A_ 3.   the second phase of a normal mixed contraction _A_
_C_ 4.   the third phase of a normal mixed contraction _E_
_D_ 5.   the fourth phase of a normal mixed contraction _D_

II.   Multiple choice

_A_ 6.   An oxygen debt in exercising skeletal muscle:
A.   is due to the buildup of lactic acid produced by anaerobic metabolism
B.   can usually be "paid back" by anaerobic metabolism
C.   occurs only when the muscle is at rest
D.   is constant, regardless of the amount of exercise
E.   has nothing to do with oxygen consumption by the skeletal muscle

_D_ 7.   Which of the following skeletal muscles have the highest ratio of motor nerve fibers to skeletal muscle fibers (i.e. motor units of the lowest number)?
A.   flexor muscles of the toes
B.   extensor muscles of the vertebral column
C.   adductor muscles of the arm
D.   flexor muscles of the fingers
E.   pronator muscles of the wrist

_B_ 8.   Approximately how many grades of contraction (degrees of control) are there in a twenty-five fiber skeletal muscle with a 1:5 motor unit arrangement?
A.   one   D.   fifteen
B.   five   E.   twenty-five
C.   ten

_D_ 9.   When a muscle contracts isotonically:
A.   the force generated increases as the length decreases
B.   the force generated is constant as the length increases
C.   the force generated decreases as the length decreases
D.   the force generated is constant as the length decreases
E.   the force generated is constant as is the length

_C_ 10.   When skeletal muscle isometrically contracts:
A.   mechanical work is performed
B.   tension developed during contraction remains constant
C.   chemical energy is converted into thermal energy (heat)
D.   its origin remains fixed but its insertion moves
E.   two of the above occur

_D_ 11.   Physiologically, the nervous system adjusts the strength of a skeletal muscle's contraction by way of:
A.   treppe   D.   motor unit recruitment
B.   tetanus   E.   contracture
C.   increasing motor unit size

_E_ 12.   The ratio between a motor nerve fiber and the number of skeletal muscle fibers it supplies is called:
A.   treppe
B.   summation
C.   reciprocal innervation
D.   tetanus
E.   a motor unit

_C_ 13.   When a skeletal muscle isometrically contracts:
A.   mechanical work is performed
B.   tension developed during contraction remains constant
C.   chemical energy is converted into thermal energy (heat)
D.   its origin remains fixed but its insertion moves
E.   two of the above occur

_A_ 14.   An oxygen debt in exercising skeletal muscle:
A.   is due to the buildup of lactic acid produced by anaerobic metabolism
B.   can usually be "paid back" by anaerobic metabolism
C.   occurs only when the muscle is at rest
D.   is constant, regardless of the amount of exercise
E.   has nothing to do with oxygen consumption by the skeletal muscle

_E_ 15.   Most skeletal muscle contractions in the body are:
A.   isotonic, not isometric
B.   isometric, not isotonic
C.   completely tetanic
D.   neither isotonic nor isometric
E.   a mix of isometric and isotonic contractions

_C_ 16. Regarding skeletal muscles, it is true that:
- A. during isotonic contraction, muscle length does not change
- B. during isometric contraction, muscle tension does not change
- C. during isometric contraction, the muscle performs no mechanical work
- D. when a muscle is resting at optimal length, it is not capable of being stimulated to perform maximal work
- E. normal contractions of skeletal muscle (such as in walking) are either isotonic or isometric but never mixed

_C_ 17. Which of the following statements is false?
- A. White muscle fibers are faster in their contractile response than red muscle fibers.
- B. White fibers fatigue more quickly than red fibers.
- C. White fibers do not function as well as red fibers under anaerobic conditions.
- D. Red fibers are more capable of producing powerful sustained contractions than are white fibers.
- E. Red fibers have more myoglobin than white fibers.

_D_ 18. Which of the following molecules can be quickly metabolized to form ATP ?
- A. Fatty acid
- B. Glucose
- C. Glycerol
- D. Creatine phosphate
- E. Lactic acid

_E_ 19. The role of the transverse tubules (t-tubules) in the excitation of skeletal muscle contraction is to:
- A. serve as a storage site for calcium ions
- B. connect the sarcomeres end to end
- C. conduct nutrients into the interior of the muscle fiber
- D. allow for the diffusion of oxygen to the interior of the muscle fiber
- E. provide an inward path for the spread of muscle action potentials

_B_ 20. The primary role of Ca^{++} in the activation of skeletal muscle is to:
- A. cause depolarization of the muscle cell plasma membrane
- B. remove the inhibition of the reaction between actin filaments and the myosin filaments
- C. activate myosin molecules so they can interact with actin filaments
- D. provide the energy necessary for contraction
- E. regulate the ionic composition of the interior of the cell.

_C_ 21. Increasing the length of a skeletal muscle prior to an isometric contraction:
- A. can only decrease the force of contraction
- B. can only increase the force of contraction
- C. may either increase or decrease the force that can be developed, depending on the length at which the muscle was originally held
- D. has no effect on the force that the muscle can develop
- E. is not possible under isometric conditions

78

NOTES

DATE : _____

*OUTLINE #*_____

QUESTIONS

1._____
2._____
3._____
4._____
5._____

OUTLINE 7

LEARNING OBJECTIVES (√)

After reading the assigned pages in the textbook, and/or reading and performing related laboratory exercises, and listening to lecture and/or laboratory presentations, you should be able to:

☐1. Outline the structural organization of the nervous system including components of the central nervous system and components of the peripheral nervous system.

☐2. Explain the classification of peripheral nerves as either sensory, motor, or mixed from the standpoint of containing afferent and/or efferent fibers.

☐3. Classify each of the 12 pairs of cranial nerves and each of the 31 pairs of spinal nerves as either sensory, motor, or mixed.

☐4. Briefly explain the general function of each of the following divisions of the nervous system: somatic motor system, somatic sensory system, special sensory system, and the autonomic nervous system.

☐5. Diagram a cross-section of the spinal cord and label each of the following: gray matter (posterior, lateral, and anterior horns), white matter (posterior, lateral, and anterior), and spinal nerve (posterior and anterior roots, and posterior root ganglion).

☐6. List the components of a spinal reflex arc and diagram a simple spinal reflex such as a withdrawal reflex.

☐7. Explain what is meant by synaptic delay, and the functional significance of synaptic delay. *old book?*

☐8. Explain why the brain receives signals from activated spinal reflex arcs but the brain's input is not essential for the initiation of a spinal reflex.

☐9. Distinguish ipsilateral spinal reflex from contralateral spinal reflex and diagram an example of each.

☐10. Explain why reflex pathways can contain inhibitory synapses as well as excitatory synapses, and diagram an example.

☐11. Draw a diagram to explain reciprocal innervation of flexors and extensors.

☐12. Diagram the myotatic reflex and explain its role in the control and coordination of skeletal muscles that produce body movement.

☐13. Explain the role of the gamma motor neuron in controlling the sensitivity of the neuromuscular spindle.

General

Reflexes

NOTES

QUESTIONS

1._____
2._____
3._____
4._____
5._____

OUTLINE 7

LECTURER:_____ NOTES_____ DATE:_____

I. GENERAL ORGANIZATION OF THE NERVOUS SYSTEM

A. Central Nervous System (CNS)
   1. Brain *- mesencephalon, pons, medulla oblingata*
   2. Spinal cord *- outer - white matter, inner gray matter*

B. Peripheral Nervous System (PNS)
   1. Fiber types
      a. Afferent (sensory) *in*

      b. Efferent (motor) *out*

   2. Cranial nerves (12 pairs) *olfactory I, optic II, oculomotor III,*
   Trigeminal V, Trochlear IV, Abducens VI, facial VII, vestibulocochler VIII
   glossopharyngeal IX, Vagus X, accessory XI, Hypoglossal XII
   3. Spinal nerves (31 pairs) *8 cervical, 12 thoracic, 5 lumbar*
   5 sacral, 1 coccygeal

C. CNS and PNS Systems
   1. Somatic motor system: control of skeletal muscle

   2. Somatic sensory system: pain, temperature, pressure, touch, conscious sense of muscle and joint movement, vibration

   3. Special sensory systems: audition (hearing), olfaction (smell), ustation (taste), equilibrium (balance) vision (sight)

   4. Visceral motor and sensory systems
      a. Sympathetic division, autonomic nervous system

      b. Parasympathetic division, autonomic nervous system

NOTES

OUTLINE #_____

QUESTIONS

1._____
2._____
3._____
4._____
5._____

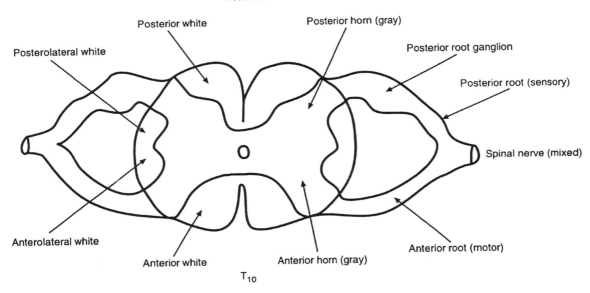

Posterior

Posterolateral white

Posterior white

Posterior horn (gray)

Posterior root ganglion

Posterior root (sensory)

Spinal nerve (mixed)

Anterolateral white

Anterior white

Anterior horn (gray)

Anterior root (motor)

T_{10}

Anterior

Cross Section, Human Spinal Cord

NOTES

DATE : _____   *OUTLINE #_____*

QUESTIONS

1._____
2._____
3._____
4._____
5._____

II. SPINAL CORD REFLEXES
A. Basic Spinal Cord Organization

1. Gray matter
   a. Posterior horn
   b. Anterior horn
   c. Lateral horn

2. White matter
   a. Posterior
   b. Lateral
   c. Anterior

3. Spinal nerve
   a. Posterior root
   b. Anterior root

B. Components of the Reflex Arc
Diagram:

C. Properties of Reflex Arcs
1. Synaptic delay

2. Relay to brain

3. Brain influence

4. Contralateral vs. ipsilateral

5. Inhibitory vs. excitatory

D. Examples of Spinal Reflexes
1. Flexion and extension
   Diagram: ipsilateral flexion-contralateral extension

NOTES

DATE : _____

OUTLINE #_____

QUESTIONS

1._____
2._____
3._____
4._____
5._____

SIMPLE SPINAL REFLEX ARC

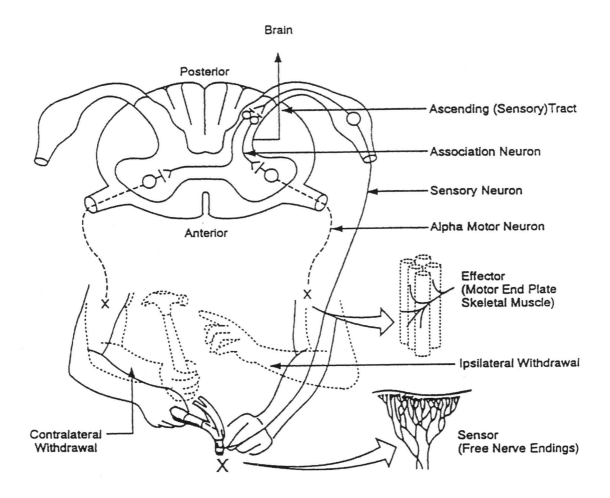

Brain

Posterior

Ascending (Sensory)Tract

Association Neuron

Sensory Neuron

Alpha Motor Neuron

Anterior

Effector
(Motor End Plate
Skeletal Muscle)

Ipsilateral Withdrawal

Contralateral
Withdrawal

Sensor
(Free Nerve Endings)

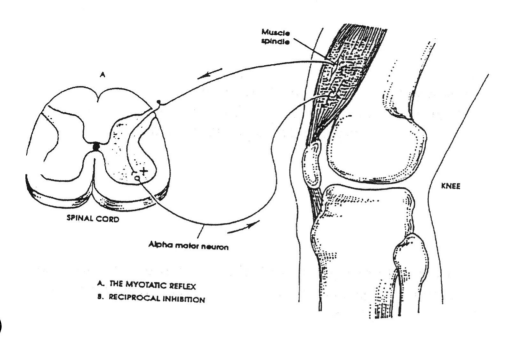

A. THE MYOTATIC REFLEX
B. RECIPROCAL INHIBITION

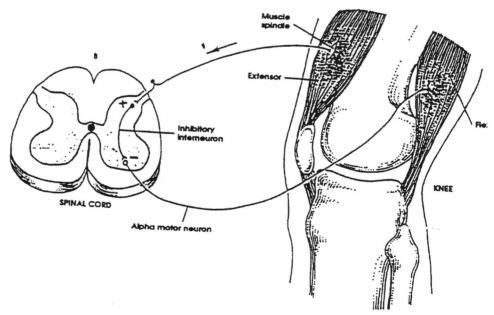

2.   Reciprocal inhibition of flexors or extensors
      Diagram: *knee-jerk reflex*

3.   The myotatic (muscle stretch) reflex
      Diagram: *a monosynaptic reflex arc; sensory neuron synapsing with alpha motor neuron*

4.   Gamma motor neuron control of neuromuscular spindle
      Diagram: *gamma motorneuron innervation of intrafusal fibers*

NOTES

QUESTIONS

1._____
2._____
3._____
4._____
5._____

MYOTATIC REFLEX
(MUSCLE STRETCH REFLEX)

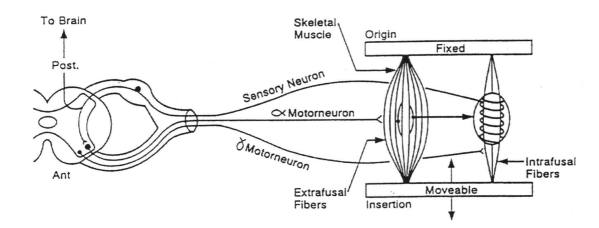

alpha

gamma

∝ Motorneurons supply extrafusal fibers to form motor units. Stimulation of motor units results in skeletal muscle contraction.

⅄ Motorneurons supply intrafusal fibers of the neuromuscular spindle. Stimulation of the gamma motorneurons increases the sensitivity to stretch of the neuromuscular spindle stretch receptor. Inhibition of the gamma motorneurons decreases the sensitivity to stretch of the neuromuscular spindle stretch receptor.

MYOTATIC REFLEX: Pull down on the moveable plate, the skeletal muscle is stretched. Since the N.M. Spindles are parallel to the extrafusal fibers, they are also stretched. As the stretch receptor is stretched it causes the sensory neuron to fire with increasing frequency thereby stimulating ∝ motorneurons to cause skeletal muscle contraction. As the skeletal muscle contracts, the moveable plate is pulled back up thereby removing the stimulus to the stretch receptor.

NOTES

QUESTIONS

1._____
2._____
3._____
4._____
5._____

gamma MN
intrafessial
X

alpa motor neuron-
Extrafussial
α

REVIEW QUESTIONS

I. Matching

A. Motor neuron
B. Association neuron
C. Sensory neuron
D. Receptor
E. Effector

D 1. the first element of a simple spinal reflex
E 2. the last element of a simple spinal reflex
C 3. the afferent fiber is connected to the soma
A 4. the efferent fiber is connected to the soma
B 5. this entire structure is within the CNS

II. Multiple choice

B 6. Stimulation of extensor gamma motorneurons:
   A. reduces stretch sensitivity of the opposing flexor
   B. increases stretch sensitivity of the extensor
   C. causes isometric contraction of the extensor
   D. inhibits the contralateral flexor
   E. none of the above

C 7. According to the principle of reciprocal innervation:
   A. extensor alpha motorneurons are stimulated when the extensor gamma motorneurons are inhibited
   B. when flexors at a joint are stimulated, extensors at the same joint are stimulated
   C. when flexor alpha motorneurons are stimulated, the alpha motor-neurons of the opposing extensor are inhibited
   D. stimulation of ipsilateral flexors is accompanied by contralateral stimulation of extensors
   E. all muscles are supplied by alpha and gamma neurons

D 8. Damage to the gamma motor neuron system would most likely result in:
   A. spastic paralysis
   B. inability to coordinate muscle activities *spinocerebellar tract*
   C. flaccid paralysis
   D. loss of myotatic reflexes
   (E.) two of the preceding

D 9. Damage to alpha motorneurons may result in:
   A. spastic paralysis of skeletal muscle
   B. paralysis of smooth muscle
   C. loss of awareness concerning muscle strength
   D. flaccid paralysis of skeletal muscle
   E. two of the preceding

_____ 10. The ipsilateral withdrawal reflex:
  A. is a monosynaptic reflex arc
  B. requires association or interneurons
  C. simultaneously activates ipsilateral extensors and flexors
  D. can easily be voluntarily inhibited once initiated
  E. two of the preceding

_____ 11. Upon stepping on a tack, a person lifts the injured foot and extends the opposite leg to maintain balance, the extensor reflex:
  A. is multisynaptic
  B. is ipsilateral
  C. requires the brain for activation
  D. is contralateral
  E. two of the preceding

_____ 12. Maintaining the forearm in a flexed position requires:
  A. inhibiting the flexor alpha motorneurons
  B. inhibiting the extensor gamma motorneurons
  C. stimulating the extensor alpha motorneurons
  D. inhibiting the flexor gamma motorneurons
  E. two of the preceding

_____ 13. The following statement is true (if more than one is true, choose E):
  A. efferent fibers in the PNS are sensory
  B. afferent fibers in the PNS are motor
  C. all cranial nerves are classified as mixed
  D. all spinal nerves are classified as mixed
  E. more than one of the above statements is true

_____ 14. Transection (completely cutting through) of the posterior roots of the spinal nerves that supply the posterior thigh would result in:
  A. inability to elicit the knee-jerk reflex
  B. spastic paralysis of posterior thigh muscles
  C. flaccid paralysis of posterior thigh muscles
  D. anesthesia of posterior thigh
  E. two of the preceding

_____ 15. Extension of the leg involves:
  A. inhibition of extensor gamma motorneurons
  B. stimulation of flexor gamma motorneurons
  C. stimulation of extensor alpha motorneurons
  D. stimulation of flexor alpha motorneurons
  E. two of the preceding

_____ 16. The primary purpose or function of the gamma motorneuron is to:
  A. prevent flexors and extensors from contracting at the same time
  B. facilitate the withdrawal reflex
  C. maintain posture of the body
  D. adjust sensitivity of the neuromuscular spindle
  E. inhibit the alpha motorneuron

_____E_____ 17.   A simple spinal reflex such as a withdrawal reflex:
  A.   is usually multisynaptic
  B.   is usually ipsilateral
  C.   usually requires input from the brain
  D.   always involves anterior and posterior spinal nerve roots
  E.   all of the above except C

_____B_____ 18.   Stimulation of extensor gamma motorneurons:
  A.   reduces stretch sensitivity of the opposing flexor
  B.   increases stretch sensitivity of the extensor
  C.   causes isometric contraction of the extensor
  D.   inhibits the contralateral flexor
  E.   none of the above

_____C_____ 19.   According to the principle of reciprocal innervation:
  A.   extensor alpha motorneurons are stimulated when the extensor gamma motorneurons are inhibited
  B.   when flexors at a joint are stimulated, extensors at the same joint are stimulated
  C.   when flexor alpha motorneurons are stimulated, the alpha motorneurons of the opposing extensor are inhibited
  D.   stimulation of ipsilateral flexors is accompanied by contralateral stimulation of extensors
  E.   all muscles are supplied by alpha and gamma neurons

_____E_____ 20.   Flexion of the forearm is accompanied by:
  A.   stimulation (+) of extensor alpha motorneurons
  B.   stimulation (+) of extensor gamma motorneurons
  C.   inhibition (-) of flexor alpha motorneurons
  D.   inhibition (-) of flexor gamma motorneurons
  E.   inhibition (-) of extensor gamma motorneurons

_____D_____ 21.   Upon stepping on a tack, a person lifts the injured foot and extends the opposite leg to maintain balance. The extensor reflex:
  A.   is monosynaptic
  B.   is ipsilateral
  C.   requires the brain for activation
  D.   is contralateral
  E.   two of the preceding

_____D_____ 22.   Which of the following motor activities best exemplifies the case where there is a simultaneous increase in gamma motorneuron activity to both agonist and antagonist muscles at a given joint?
  A.   walking
  B.   running
  C.   swimming
  D.   standing at attention
  E.   lifting bar-bells

_B__ 23. Gamma motor neurons:
  A. are found in the dorsal root of a spinal nerve
  B. stimulate intrafusal fibers
  C. inhibit extrafusal fibers
  D. are controlled by extra pyramidal tracts
  E. two of the preceding

Pg 318 + 19

_D__ 24. Transection (completely cutting through) of the anterior roots of the spinal nerves that supply the muscles of the posterior thigh would result in:
  A. inability to voluntarily extend the leg
  B. inability to elicit the knee-jerk reflex
  C. spastic paralysis of posterior thigh muscles
  D. flaccid paralysis of posterior thigh muscles
  E. two of the preceding

24

_E__ 25. Alpha motorneurons:
  A. are found in the posterior root of a spinal nerve
  B. are controlled by upper motorneurons
  C. are lower motorneurons
  D. adjust sensitivity of the neuromuscular spindle gamma
  E. two of the preceding

Pg 310
334

_E__ 26. (C) Which of the following reflex arcs contains no association neuron?
  A. withdrawal from painful stimulus
  B. contralateral extension coupled to reflex
  C. knee-jerk elicited by patellar hammer
  D. reciprocal inhibition
  E. more than one of the above

_D__ 27. A time when you would want to simultaneously increase spindle sensitivity in both flexors and extensors at a given joint:
  A. when walking
  B. when running
  C. when swimming
  D. when standing at attention
  E. there never is such a time

_D__ 28. Damage to the alpha motor neurons may result in:
  A. spastic paralysis of skeletal muscle
  B. paralysis of smooth muscle
  C. loss of awareness concerning muscle strength
  D. flaccid paralysis of skeletal muscle
  E. increased tonus of the muscle

28

decussation - crossover

OUTLINE 8

LEARNING OBJECTIVES (√)

After reading the assigned pages in the textbook, and/or reading and performing related laboratory exercises, and listening to lecture and/or laboratory presentations, you should be able to:

□1. Define spinal cord tract and distinguish between ascending, descending, and intersegmental tracts.

□2. For each of the following sensory tracts; posterior columns, anterior and lateral spinothalamic tracts, anterior and posterior spinocerebellar tracts, answer the following questions:
   a.) What kind of sensory information does the tract carry?
   b.) How many neurons (1st order, 2nd order, etc.) are there in the pathway and where are they located?
   c.) Does the information become contralateral, and if so, where?
   d.) Where does the tract terminate?

□3. For each of the following motor tracts; lateral corticospinal tracts, anterior corticospinal tracts, answer the following questions:
   a.) Where does the tract begin and where does the tract terminate?
   b.) Describe the motor control functions of each tract.
   c.) How many neurons are there in the pathway, where are they located, and what are their names?

□4. Explain major functional differences between the corticospinal tracts and the extracorticospinal tracts.

□5. Explain the loss of sensation (anaesthesia) and the loss of somatic motor control (paralysis, loss of coordination, etc.) that may occur as a result of a lesion in the spinal nerve, spinal cord, brainstem, thalamus, or cerebrum.

□6. Use a mushroom analogy to describe the general anatomic organization of the brain with reference to the relation of the following parts: medulla oblongata, pons, mesencephalon, thalamus, hypothalamus, cerebrum, and cerebellum.

□7. Give the location of the nuclei and list the specific functions of each of the following cranial nerves: I. Olfactory, II. Optic, III. Oculomotor, IV. Trochlear, V. Trigeminal, VI. Abducens, VII. Facial, VIII. Vestibulocochlear, IX. Glossopharyngeal, X. Vagus, XI. Accessory, and XII. Hypoglossal.

□8. Describe the motor and/or sensory loss that may occur when each of the cranial nerves is unilaterally damaged.

NOTES

DATE : _____   *OUTLINE #_____*

QUESTIONS

1._____
2._____
3._____
4._____
5._____

OUTLINE 8

LECTURER:_____   NOTES_____   DATE:_____

I.   **TRACTS OF THE SPINAL CORD**
    A.   Ascending Tracts (Sensory)
        1.   Posterior columns (dorsal columns)
           Diagram appended
        2.   Anterior spinothalamic tracts
           Diagram appended
        3.   Lateral spinothalamic tracts
           Diagram appended
        4.   Spinocerebellar tracts (anterior & posterior)
           Diagram appended

    B.   Descending Tracts (Motor)
        1.   Lateral corticospinal tracts
           Diagram appended
        2.   Anterior corticospinal tracts
           Diagram appended
        3.   Extracorticospinal tracts
           a.   rubrospinal tract
           b.   tectospinal tract
           c.   vestibulospinal tract
           d.   olivospinal tract
           e.   reticulospinal tract

    C.   Spinal Cord Damage
        1.   Loss of sensation
           Example: hemisection at T_6
        2.   Loss of motor control
           Example: hemisection at T_6

II. **ORGANIZATION OF THE BRAIN** (diagram attached)
    A.   Brainstem   **B.**   Diencephalon   **C.** Cerebellum
        1.   Medulla oblongata   1.   Thalamus   **D.** Cerebrum
        2.   Pons   2.   Hypothalamus
        3.   Mesencephalon

NOTES

DATE : _____   *OUTLINE #_____*

QUESTIONS

1._____
2._____
3._____
4._____
5._____

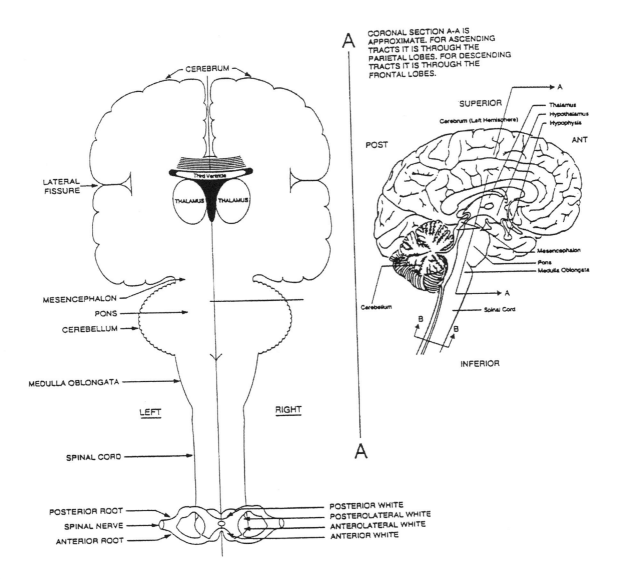

CORONAL SECTION A-A IS
APPROXIMATE. FOR ASCENDING
TRACTS IT IS THROUGH THE
PARIETAL LOBES. FOR DESCENDING
TRACTS IT IS THROUGH THE
FRONTAL LOBES.

CEREBRUM

LATERAL
FISSURE

Third Ventricle

THALAMUS   THALAMUS

MESENCEPHALON
PONS
CEREBELLUM

MEDULLA OBLONGATA

LEFT   RIGHT

SPINAL CORD

POSTERIOR ROOT
SPINAL NERVE
ANTERIOR ROOT

POSTERIOR WHITE
POSTEROLATERAL WHITE
ANTEROLATERAL WHITE
ANTERIOR WHITE

A

A

SUPERIOR

POST   ANT

Thalamus
Hypothalamus
Hypophysis
Cerebrum (Left Hemisphere)

Mesencephalon
Pons
Medulla Oblongata

Cerebellum

A

B   B

Spinal Cord

INFERIOR

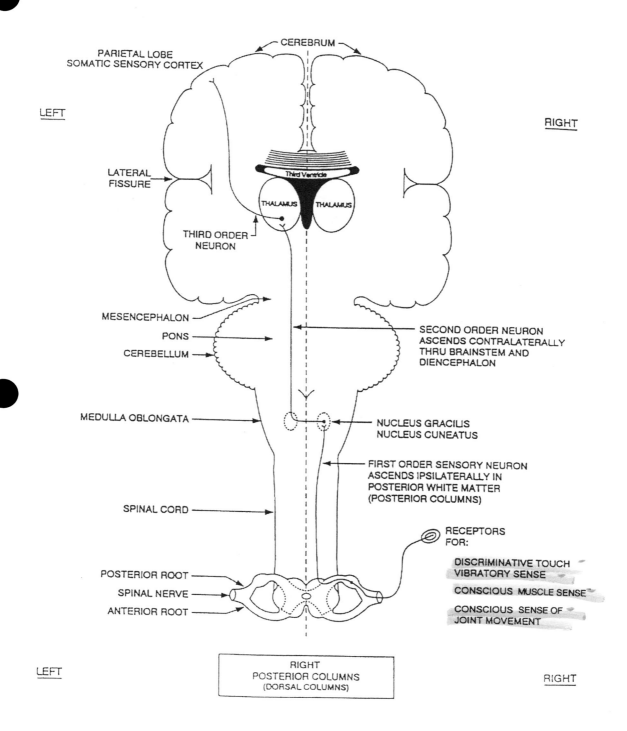

CEREBRUM

PARIETAL LOBE
SOMATIC SENSORY CORTEX

LEFT

RIGHT

Third Ventricle

THALAMUS   THALAMUS

LATERAL
FISSURE

THIRD ORDER
NEURON

MESENCEPHALON

PONS

CEREBELLUM

SECOND ORDER NEURON
ASCENDS CONTRALATERALLY
THRU BRAINSTEM AND
DIENCEPHALON

MEDULLA OBLONGATA

NUCLEUS GRACILIS
NUCLEUS CUNEATUS

FIRST ORDER SENSORY NEURON
ASCENDS IPSILATERALLY IN
POSTERIOR WHITE MATTER
(POSTERIOR COLUMNS)

SPINAL CORD

RECEPTORS
FOR:

DISCRIMINATIVE TOUCH
VIBRATORY SENSE

POSTERIOR ROOT

SPINAL NERVE

CONSCIOUS MUSCLE SENSE

ANTERIOR ROOT

CONSCIOUS SENSE OF
JOINT MOVEMENT

LEFT

RIGHT
POSTERIOR COLUMNS
(DORSAL COLUMNS)

RIGHT

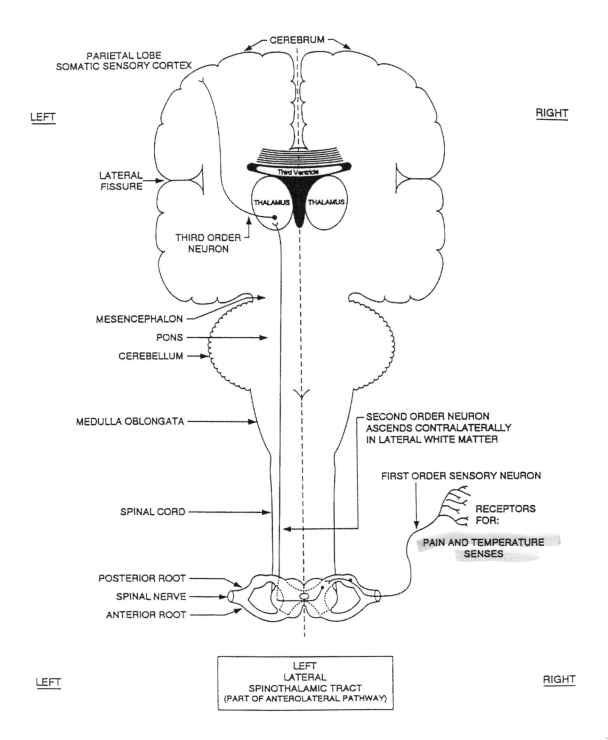

CEREBRUM

PARIETAL LOBE
SOMATIC SENSORY CORTEX

LEFT

RIGHT

Third Ventricle

LATERAL
FISSURE

THALAMUS   THALAMUS

THIRD ORDER
NEURON

MESENCEPHALON

PONS

CEREBELLUM

SECOND ORDER NEURON
ASCENDS CONTRALATERALLY
IN LATERAL WHITE MATTER

MEDULLA OBLONGATA

FIRST ORDER SENSORY NEURON

RECEPTORS
FOR:

SPINAL CORD

PAIN AND TEMPERATURE
SENSES

POSTERIOR ROOT

SPINAL NERVE

ANTERIOR ROOT

LEFT

LEFT
LATERAL
SPINOTHALAMIC TRACT
(PART OF ANTEROLATERAL PATHWAY)

RIGHT

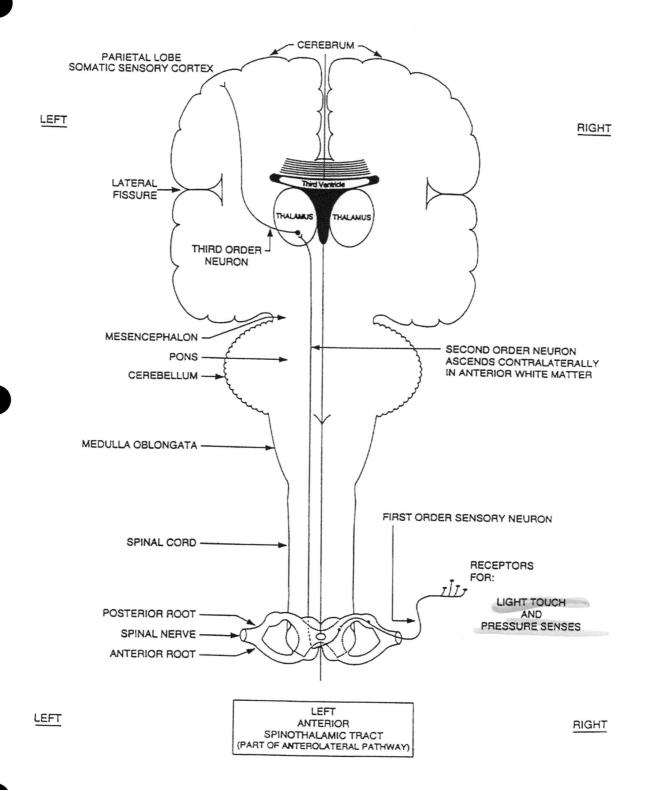

CEREBRUM

PARIETAL LOBE
SOMATIC SENSORY CORTEX

LEFT

RIGHT

LATERAL
FISSURE

Third Ventricle

THALAMUS   THALAMUS

THIRD ORDER
NEURON

MESENCEPHALON

PONS

CEREBELLUM

SECOND ORDER NEURON
ASCENDS CONTRALATERALLY
IN ANTERIOR WHITE MATTER

MEDULLA OBLONGATA

FIRST ORDER SENSORY NEURON

SPINAL CORD

RECEPTORS
FOR:

POSTERIOR ROOT

SPINAL NERVE

ANTERIOR ROOT

LIGHT TOUCH
AND
PRESSURE SENSES

LEFT

RIGHT

LEFT
ANTERIOR
SPINOTHALAMIC TRACT
(PART OF ANTEROLATERAL PATHWAY)

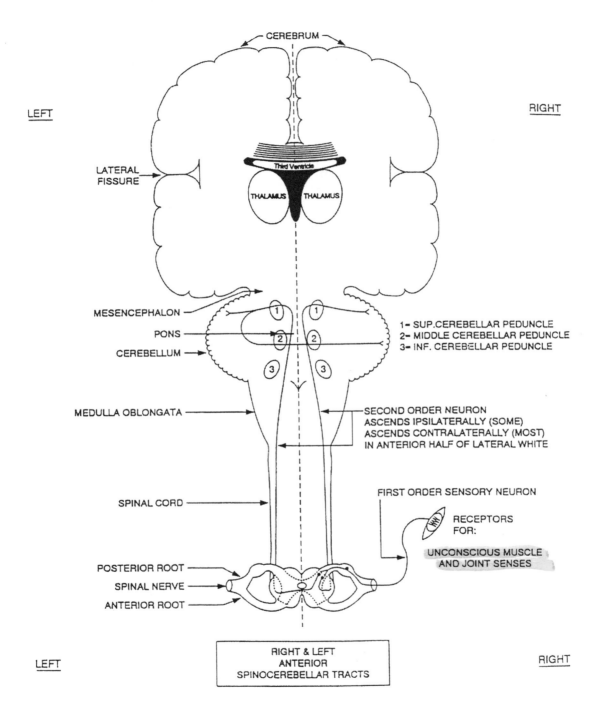

CEREBRUM

LEFT

RIGHT

LATERAL
FISSURE

Third Ventricle

THALAMUS   THALAMUS

MESENCEPHALON

1

1

PONS

2

2

1= SUP. CEREBELLAR PEDUNCLE
2= MIDDLE CEREBELLAR PEDUNCLE
3= INF. CEREBELLAR PEDUNCLE

CEREBELLUM

3

3

MEDULLA OBLONGATA

SECOND ORDER NEURON
ASCENDS IPSILATERALLY (SOME)
ASCENDS CONTRALATERALLY (MOST)
IN ANTERIOR HALF OF LATERAL WHITE

FIRST ORDER SENSORY NEURON

SPINAL CORD

RECEPTORS
FOR:

UNCONSCIOUS MUSCLE
AND JOINT SENSES

POSTERIOR ROOT
SPINAL NERVE
ANTERIOR ROOT

RIGHT & LEFT
ANTERIOR
SPINOCEREBELLAR TRACTS

LEFT

RIGHT

106

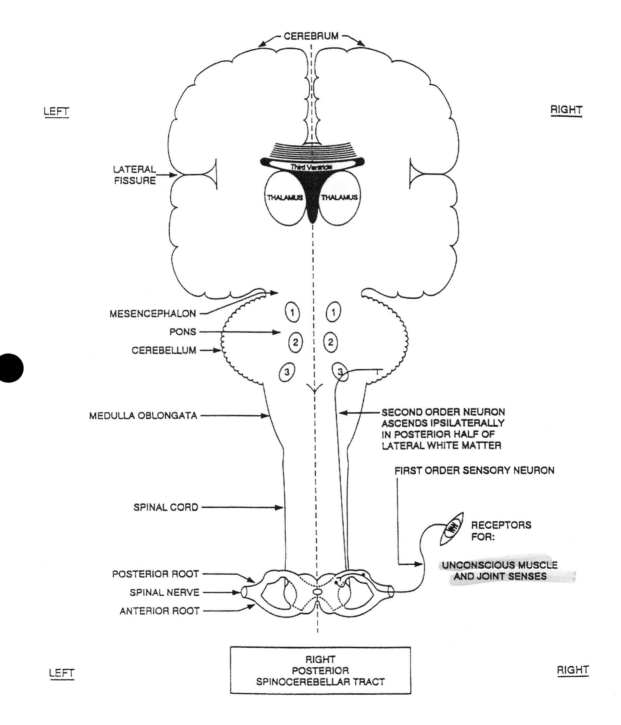

CEREBRUM

LEFT

RIGHT

LATERAL
FISSURE

Third Ventricle

THALAMUS   THALAMUS

MESENCEPHALON

PONS

CEREBELLUM

MEDULLA OBLONGATA

SPINAL CORD

POSTERIOR ROOT

SPINAL NERVE

ANTERIOR ROOT

SECOND ORDER NEURON
ASCENDS IPSILATERALLY
IN POSTERIOR HALF OF
LATERAL WHITE MATTER

FIRST ORDER SENSORY NEURON

RECEPTORS
FOR:

UNCONSCIOUS MUSCLE
AND JOINT SENSES

LEFT

RIGHT

RIGHT
POSTERIOR
SPINOCEREBELLAR TRACT

107

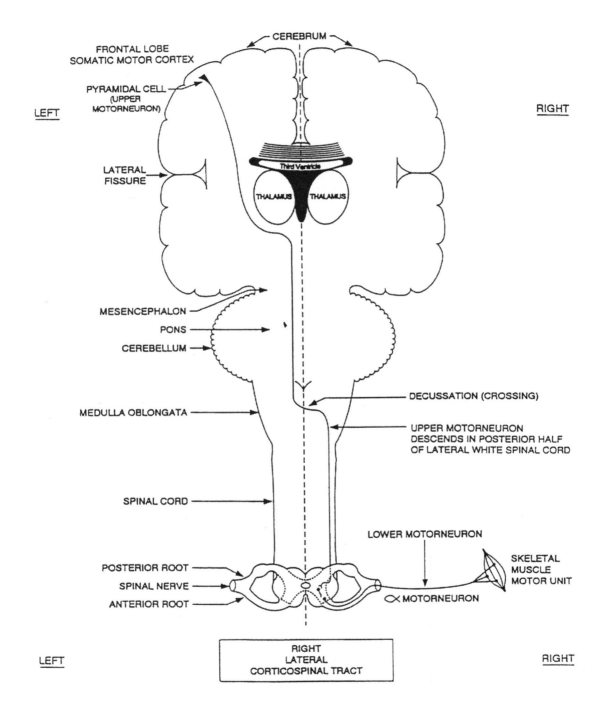

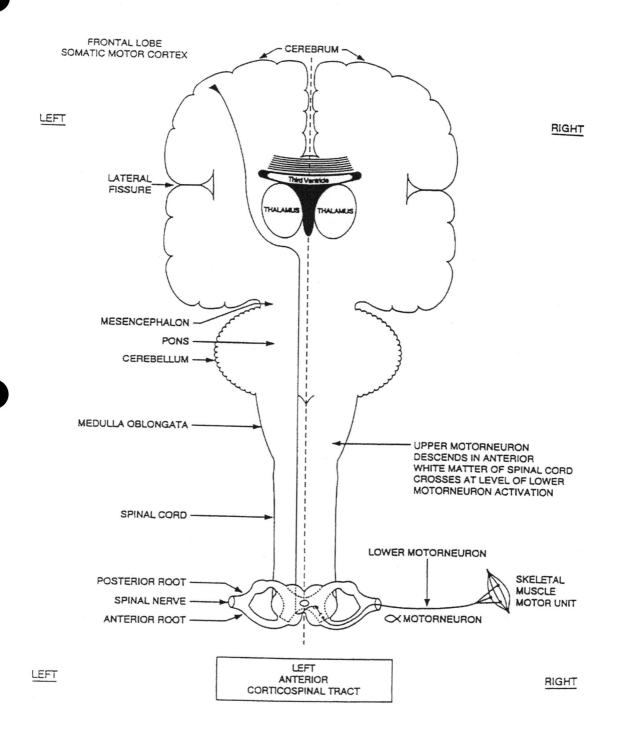

FRONTAL LOBE
SOMATIC MOTOR CORTEX

CEREBRUM

LEFT

RIGHT

Third Ventricle

THALAMUS   THALAMUS

LATERAL
FISSURE

MESENCEPHALON

PONS

CEREBELLUM

MEDULLA OBLONGATA

UPPER MOTORNEURON
DESCENDS IN ANTERIOR
WHITE MATTER OF SPINAL CORD
CROSSES AT LEVEL OF LOWER
MOTORNEURON ACTIVATION

SPINAL CORD

LOWER MOTORNEURON

POSTERIOR ROOT

SPINAL NERVE

ANTERIOR ROOT

SKELETAL
MUSCLE
MOTOR UNIT

∝ MOTORNEURON

LEFT

LEFT
ANTERIOR
CORTICOSPINAL TRACT

RIGHT

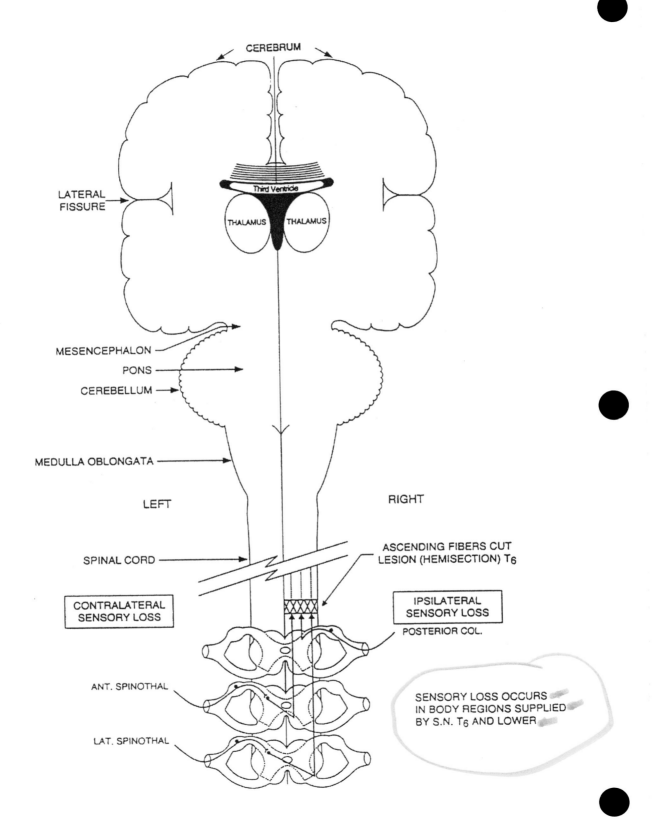

CEREBRUM

LATERAL FISSURE

Third Ventricle

THALAMUS   THALAMUS

MESENCEPHALON

PONS

CEREBELLUM

MEDULLA OBLONGATA

LEFT

RIGHT

SPINAL CORD

ASCENDING FIBERS CUT
LESION (HEMISECTION) T6

CONTRALATERAL
SENSORY LOSS

IPSILATERAL
SENSORY LOSS

POSTERIOR COL.

ANT. SPINOTHAL

SENSORY LOSS OCCURS
IN BODY REGIONS SUPPLIED
BY S.N. T6 AND LOWER

LAT. SPINOTHAL

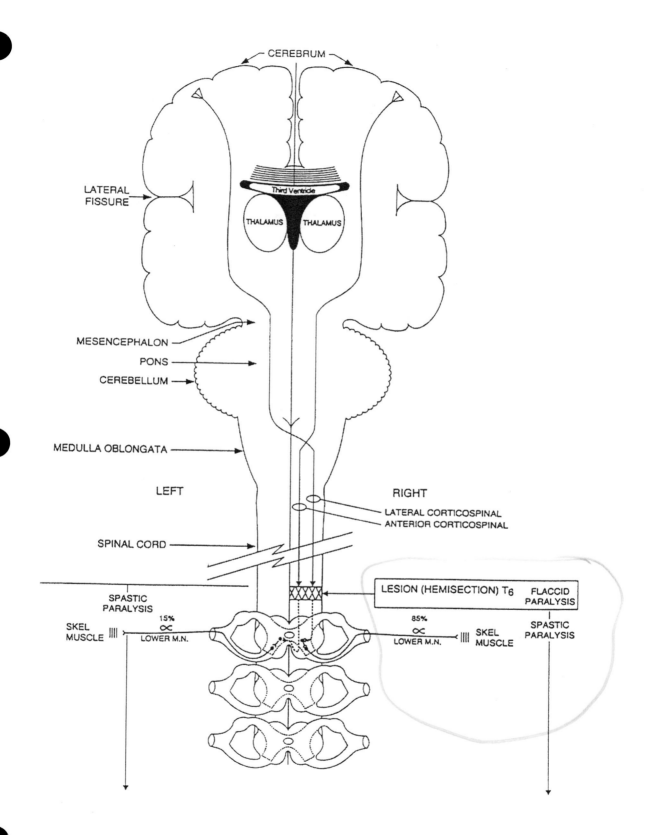

CEREBRUM

LATERAL
FISSURE

Third Ventricle

THALAMUS   THALAMUS

MESENCEPHALON

PONS

CEREBELLUM

MEDULLA OBLONGATA

LEFT   RIGHT

LATERAL CORTICOSPINAL
ANTERIOR CORTICOSPINAL

SPINAL CORD

SPASTIC   LESION (HEMISECTION) T6   FLACCID
PARALYSIS   PARALYSIS

SKEL   15%   85%   SPASTIC
MUSCLE   ∝   ∝   SKEL   PARALYSIS
          LOWER M.N.   LOWER M.N.   MUSCLE

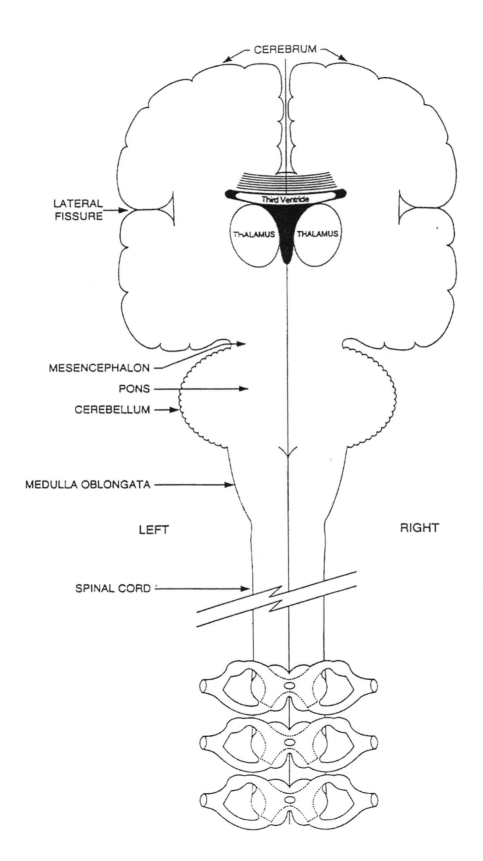

CEREBRUM

LATERAL
FISSURE

Third Ventricle

THALAMUS   THALAMUS

MESENCEPHALON

PONS

CEREBELLUM

MEDULLA OBLONGATA

LEFT   RIGHT

SPINAL CORD

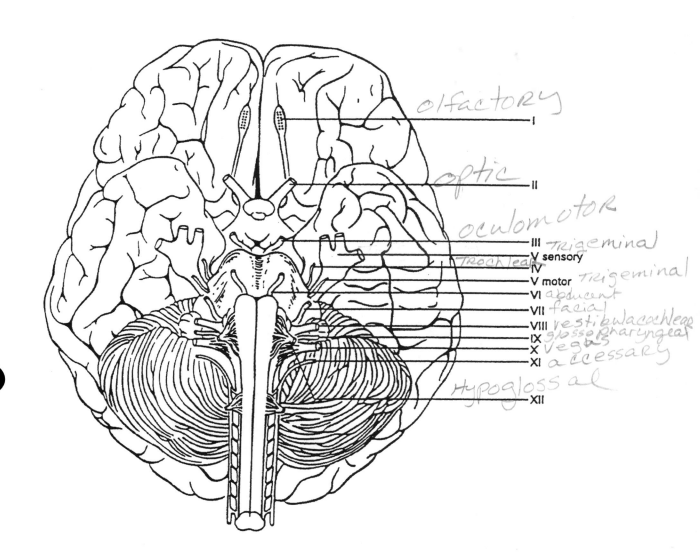

olfactory
I

optic
II

oculomotor
III TRIGEMINAL
V sensory
TROchleaR
IV
V motor TRigeminal
VI abducent
VII facial
VIII vestibulacochleae
IX glossopharyngeal
X vegas
XI accessary

Hypoglossal
XII

SUMMARY OF CRANIAL NERVES

Cranial Nerve	Distribution and Function
I. Olfactory (sensory)	Sensory from Nasal epithelium. Nerve fibers pass through Cribiform Plate into Olfactory Bulb. Fibers pass from Bulb back along Olfactory Tract to Cerebrum.
II. Optic (sensory)	Sensory from Retina. Optic Nerve arises from cells in Retina. Nasal half of fibers distributed to opposite side of brain (cross at Chiasma). Fibers distributed to Brain Stem and Occipital Lobe of Cortex.
III. Oculomotor (motor)	Arises from undersurface of Brain Stem in front of Pons. Passes forward into the Orbit and supplies Sup. Med. and Inf. Rectus muscles and Inf. Oblique muscles. Also motor to the smooth muscle of Iris and Ciliary body (Parasympathetic).
IV. Trochlear (motor)	Arises from Dorsal surface of Midbrain, passes around side of Brain Stem and forward into the Orbit. Supplies Sup Oblique muscle.
V. Trigeminal (mixed)	Has 2 roots, sensory & Motor, arising from side of Pons. Sensory root (heat, cold, pain, touch) — (1) Ophthalmic Nerve: eyes, nose, front scalp (2) Maxillary Nerve: lower eyelid, nose, upper cheek, lip, jaw and palate (3) Mandibular Nerve: lower lip, jaw, tongue, face, cheek, ear. Motor root—Muscles of mastication, some neck muscles
VI. Abducent (motor)	Arises from Brain Stem just behind Pons and passes forward into the Orbit, motor to Lateral Rectus muscle.
VII. Facial (mixed)	Arises from Brain Stem just behind Pons, lateral to Abducent Nerve. Motor — Muscles of facial expression. Parasympathetic Innervation: A few neck muscles, Submandibular Glands, Sublingual Glands and Lacrimal Gland. Sensory — Taste from Anterior 2/3 tongue. Sensory from Facial muscle (Proprioceptive)
VIII. Vestibulocochlear (sensory)	Arises from Brain Stem lateral to Facial nerve. Sensory from Inner Ear.
IX. Glossopharyngeal (mixed)	Arises from side of Medulla and passes out of skull via Jugular Foramen. Motor — Secretory fibers to Parotid Glands, supplies Pharyngeal muscles, muscles of Larynx and Soft Palate via Pharyngeal Plexus (9-10-11). Sensory — From Pharynx, Taste from post. 1/3 tongue, from Carotid Sinus (blood pressure and chemistry)
X. Vagus (mixed)	Arises from side of Medulla just behind 9th C.N. and passes out of skull via the Jugular Foramen. Motor — Muscles of Pharynx, Larynx (Pharyngeal Plexus), Esophagus, Heart, Lungs, Stomach, Intestines. Sensory— From Laryngeal Mucosa, Heart, Lungs, Bronchi, Esophagus, Stomach, Intestines
XI. Accessory (motor)	Arises from side of Medulla below X passes out of skull via Jugular Foramen and Foramen Magnum. Motor to muscles of Larynx and Pharynx (Pharyn. Plexus) motor to Sternocleidomastoid and Trapezius muscles.
XII. Hypoglossal (motor)	Arises from side of Medulla below XI passes out of skull via Hypoglossal Canal. Supplies muscles of Tongue.

Handwritten annotations: Encephalon; para; Eye movement 4 muscles; Pons; para; more nucleus; Hearing; para; seros - salivation for talking; Parasympathetic; nerves 9,10,11 - speech swallowing; not salivary; Medulla oblongata

nerves 9, 10, 11 - speech, swallowing,

III. CRANIAL NERVES: DISTRIBUTION AND FUNCTION
(see attached list of functions and diagram of anterior or ventral brain)

A. Cranial Nerve I: Olfactory - *sensory - nasal*

B. Cranial Nerve II: Optic - *sensory - eyes*

C. Cranial Nerve III: Oculomotor - *motor - superior, medial and inferior Rectus muscles, and inferior oblique + smooth muscles of iris + ciliary body (parasympathetic)*

D. Cranial Nerve IV: Trochlear *motor - super oblique - muscle*

E. Cranial Nerve V: Trigeminal - *mixed - 2 roots - sensory ophthalamic + maxillary + mandibular nerves motor - mastication + neck muscles*

F. Cranial Nerve VI: Abducens - *motor - lateral Rectus -*

G. Cranial Nerve VII: Facial - *mixed - motor - facial expression - mucous secretions ← submandible glands, sublingual and lacrimal glands sensory - 2/3 anterior tongue proprioceptive*

H. Cranial Nerve VIII: Vestibulocochlear - *sensory from inner ear*

I. Cranial Nerve IX: Glossopharyngeal - *motor - Partid glands - serous fluid for talking - muscles of larynx + soft palate carotid sinus sensory - from Pharynx, taste from posterior 1/3 of tongue*

J. Cranial Nerve X: Vagus - *motor - muscles - pharynx, larynx, esoph. Heart lungs, stomach, Intestines* mix parasymp. *sensory - larynegeal mucos - Heart, lungs, bronchi, esopha. stomach, intestines*

K. Cranial Nerve XI: Spinal Accessory - *muscles of larynx, pharynx* motor *Sternocleidomastoid & trapezius*

L. Cranial Nerve XII: Hypoglossal - motor *tongue*

NOTES

DATE : _____   *OUTLINE #*_____

QUESTIONS

1._____
2._____
3._____
4._____
5._____

REVIEW QUESTIONS

I. Matching

A.   Facial Nerve
B.   Trigeminal nerve
C.   Optic nerve
D.   Oculomotor nerve
E.   Glossopharyngeal nerve

_B__ 1.   Motor nerve to muscles of mastication *B*
_D__ 2.   Motor nerve to pupillary constrictor muscle *D*
_A__ 3.   Sensory (pain, temperature, touch) from the skin of the face *A*
_E__ 4.   Sensory from taste buds *E*
_C__ 5.   Sensory from the retina *C*

II. Multiple choice

B 6.   If the sensory portion of this nerve were cut, facial anesthesia would result:
   A.   C.N. III
   B.   C.N. V
   C.   C.N. II
   D.   C.N. IX
   E.   C.N. VI

E 7.   The ability to discriminate between two mechanical stimuli applied to skin requires normal functioning of the:
   A.   anterior spinothalamic tracts
   B.   ventral spinocerebellar tracts
   C.   dorsal columns
   D.   parietal cortex of cerebrum
   E.   two of the preceding

C 8.   Rotation of the eyeball in the orbit requires cranial nerves _____.
   A.   5-6-7
   B.   2-3-5
   C.   3-4-6
   D.   2-4-6
   E.   3-4-5

D 9.   A loss of taste and salivation could result from damage to the _____ cranial nerve.
   A.   fifth
   B.   twelfth
   C.   second
   D.   seventh
   E.   none of the preceding

_D_ 10.   If your spinal cord was completely severed between C_3 and C_4 vertebrae you would lose:
    A.   pain sensation from the face
    B.   myotatic reflexes of the lower extremities
    C.   myotatic reflexes of the upper extremities
    D.   upper motorneuron control of respiratory muscles
    E.   reflex emptying of the urinary bladder

_D_ 11.   Which of the following tracts transmits pain information from the skin on the left thigh?
    A.   left anterior spinothalamic
    B.   right anterior spinothalamic
    C.   left lateral spinothalamic
    D.   right lateral spinothalamic
    E.   left posterior column

_C_ 12.   Damage to the following nerve may result in paralysis of facial muscles:
    A.   C.N. III oculacmotor
    B.   C.N. V trigeminal
    C.   C.N. VII facial
    D.   C.N. X vagus
    E.   C.N. VI abducent

_A_ 13.   Which nerve controls heart rate, gastric secretion, and pancreatic secretion?
    A.   C.N. X vagus
    B.   C.N. IX glossopharyngeal
    C.   C.N. VII facial
    D.   C.N. XII Hypoglossal
    E.   C.N. XI accessory

_D_ 14. (A) Deviation of the tongue to the left suggests damage to the:
    A.   right C.N. XII Hypoglossal
    B.   left C.N. V trigeminal
    C.   left C.N. IX glossopharyngeal
    D.   left C.N. XII Hypoglossal
    E.   left C. N. VII facial

_C_ 15.   The motor division of cranial nerve V on the right side of the brainstem has become nonfunctional. As a result, the subject has lost entirely or partially the ability to:
    A.   raise the right eyebrow
    B.   frown
    C.   clench teeth
    D.   sense pain from the right cheek skin
    E.   shed tears

_C_ 16. Damage to the lateral spinothalamic pathway in the left brainstem could result in:
    A.   loss of cutaneous pain on the left side of the trunk
    B.   loss of light touch from the left arm skin
    C.   loss of cutaneous temperature sensation from the right leg
    D.   dilated left pupil
    E.   spastic paralysis of the arm

_E_ 17. _(A)_ Damage to the dorsal columns, such as may occur in tabes dorsalis (syphilis) would result in partial or complete:
    A.   loss of cutaneous two point discrimination
    B.   loss of skeletal muscle coordination
    C.   loss of pain sensation
    D.   loss of temperature sensation
    E.   answers (A) and (B)

_A_ 18. _(C)_ Damage to the lateral white matter of the spinal cord on the right side of the body between the shoulder blades could result in:

P 268

    A.   loss of spinal reflexes involving the left arm
    B.   loss of spinal reflexes involving the right leg
    C.   loss of sensation from the skin of the left leg
    D.   paralysis of the right side of the face
    E.   sensory and motor loss involving the left leg

_D_ 19. Damage to cranial nerve V would most likely interfere with:
    A.   vision
    B.   hearing
    C.   swallowing
    D.   chewing
    E.   balance

_E_ 20. _(A)_ Damage to the dorsal columns, such as may occur in tabes dorsalis (syphilis) would result in partial or complete:
    A.   loss of vibratory sensation
    B.   loss of skeletal muscle coordination
    C.   loss of pain sensation
    D.   loss of temperature sensation
    E.   answers (A) and (B)

_B_ 21. The right corner of the mouth and the right upper eyelid droop. There is noticeable lack of muscle tone in the right cheek. This person suffers from damage to:
    A.   left C.N. V trigeminal
    B.   right C.N. VII facial
    C.   right C.N. III oculomotor
    D.   left C.N. IX glossopharyngeal
    E.   right C.N. VI abducent

119

22. A patient involved in a sailing accident has lost the ability to sense the flow of air across his left arm and leg and to perceive pain on the right side of his body. These symptoms are suggestive of:

A. a hemisection of the right side of the brain stem above the level of the medulla

B. a hemisection of the left side of the spinal cord at the cervical level

C. a hemisection of the right side of the spinal cord at the thoracic level

D. a hemisection of the left side of the spinal cord at the thoracic level

E. none of the preceding

23. Which of the following cranial nerves supply motor fibers to muscles that raise the eyebrows?

A. cranial nerve II optic

B. cranial nerve V trigeminal

C. cranial nerve VI abducent

D. cranial nerve VII facial

E. cranial nerve IX glossopharyngeal

24. Damage to this cranial nerve impairs salivation:

A. facial nerve

B. trigeminal nerve

C. glossopharyngeal nerve   salivation For talking

D. vagus nerve

E. spinal accessory nerve

25. If transmission of sensory information in this tract were blocked, the patient would be unable to sense pain from skin on the left side of the body:

A. left anterior spinothalamic tract

B. right lateral spinothalamic tract

C. right posterior column

D. left lateral spinothalamic tract

E. right anterior spinothalamic tract

26. Accommodation of the lens requires normal function of:

A. the optic nerves II

B. the oculomotor nerves III

C. the facial nerves VII

D. the trigeminal nerves V

E. both A and B

120

NOTES

QUESTIONS

1._____
2._____
3._____
4._____
5._____

OUTLINE 9

LEARNING OBJECTIVES (√)

After reading the assigned pages in the textbook, and/or reading and performing related laboratory exercises, and listening to lecture and/or laboratory presentations, you should be able to:

☐1.　List the cranial nerve nuclei located in the medulla oblongata and briefly outline the functions controlled by this part of the brainstem.

☐2.　List the cranial nerve nuclei located in the pons and briefly outline the functions controlled by this part of the brainstem.

☐3.　List the cranial nerve nuclei located in the mesencephalon and briefly outline the functions controlled by this part of the brainstem.

☐4.　Describe three major functions of the thalamus.

☐5.　Identify five physiological processes controlled or regulated by nuclei in the hypothalamus.

☐6.　Define gyrus, sulcus, fissure, cortex, and medulla of the cerebrum.

☐7.　Identify the following fissures/sulci and explain how each anatomical landmark is used to divide the cerebrum into hemispheres and/or lobes: longitudinal fissure, central fissure (cruciate fissure), lateral fissure, parieto-occipital sulcus.

☐8.　Define precentral gyrus, primary somatic motor cortex, and motor homunculus.

☐9.　Outline the function of the primary somatic motor cortex and give an example of loss when this area of the brain has been damaged.

☐10.　Locate the premotor cortex and briefly explain its control of motor activities, including programmed movement and conjugate eye movement.

☐11.　Define Broca's area and indicate hemispheric dominance.

☐12.　Explain the role of the prefrontal cortex in integration, synthesis, and behavior, and give an example of behavioral change associated with a lesion in this area.

☐13.　Define postcentral gyrus, primary somatic sensory cortex, and sensory homunculus.

☐14.　Outline the function of the primary somatic sensory cortex and give an example of loss when this area of the brain has been damaged.

☐15.　Explain the functions of the parietal somaesthetic association cortex and give an example of loss when this area of the brain has been damaged.

☐16.　Outline the functions of the primary and association cortices of the temporal and the occipital lobes. Explain the effect of a lesion in each area.

☐17.　Explain the role of the cerebellum in the control and integration of somatic motor activities and in vestibular and postural reflexes.

● cranial nerve nuclei: Nerve V - spinal tract nucleus
Pg. 268

cranial nerves 209 + 341

nucleus cutaneous 265
nucleus gracilis 265

cranial nerve nuclei - PONS

mesencephalon

● Thalmus - 3 functions

cerebellum
1) planning movement
2) controls posture
3) control smooth limb movement

3 lobes
1) anterior
2) posterior
3) flocculonodular

QUESTIONS

1._____
2._____
3._____
4._____
● 5._____

OUTLINE 9

I.   **BRAINSTEM FUNCTIONS**
    A.   Medulla Oblongata: C.N. nuclei 9-10-11-12

    B.   Pons: C.N. nuclei 5-6-7-8

    C.   Mesencephalon: C.N.nuclei 2-3-4

II. **DIENCEPHALON FUNCTIONS**
    A.   Thalamus
        1.   Somatic sensory relay
        2.   Special sensory relay
        3.   Sensory integration

    B.   Hypothalamus
        1.   Autonomic control centers
        2.   Endocrine control centers

III. **CEREBRUM**
    A.   General Structure: Gyri, Sulci, Lobes, Cortex, Medulla
        Diagrams: appended
    B.   Frontal Lobe
        1.   Primary motor and premotor cortices

        2.   Broca's area

        3.   Prefrontal cortex: integration, synthesis, behavior

    C.   Parietal Lobe
        1.   Primary somatic sensory cortex

        2.   Somesthetic association cortex

NOTES

DATE : _____   *OUTLINE #_____*

QUESTIONS

1._____
2._____
3._____
4._____
5._____

D. Temporal Lobe
   1. Primary auditory cortex

   2. Auditory association cortex

E. Occipital Lobe
   1. Primary visual cortex

   2. Visual association cortex

IV. **CEREBELLUM** *- plan, control, cordination, Posture*
   A. Control and Integration: Somatic Motor Activities
      Example: cerebellar lesions and abnormal motor control

Predominate cell - Purkinje Cell
GABA = Hyperpolarization
+ inhibition

   B. Vestibular and Postural Reflexes
      Diagram: control of myotatic reflexes

Flocculonodular Lobe = equilibrium + posture

Anterior Lobe + posterior = midline + intermediate = limb movement

Anterior + posterior lobes - Lateral = planning + initiating

NOTES

cerebellar nuclei - ④ denate
interpositus = globase & emboliform
fastigial

QUESTIONS

1._____
2._____
3._____
4._____
5._____

Cerebrum (Left Hemisphere)

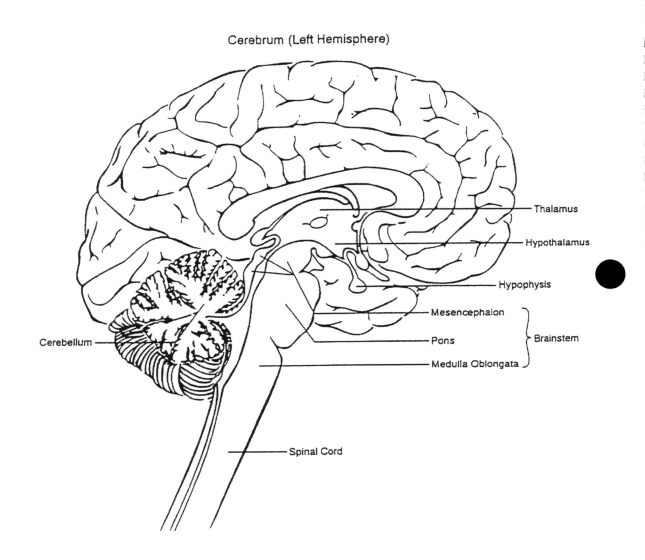

Thalamus

Hypothalamus

Hypophysis

Mesencephalon

Pons

Medulla Oblongata

Brainstem

Cerebellum

Spinal Cord

128

Cerebrum (Right Hemisphere)

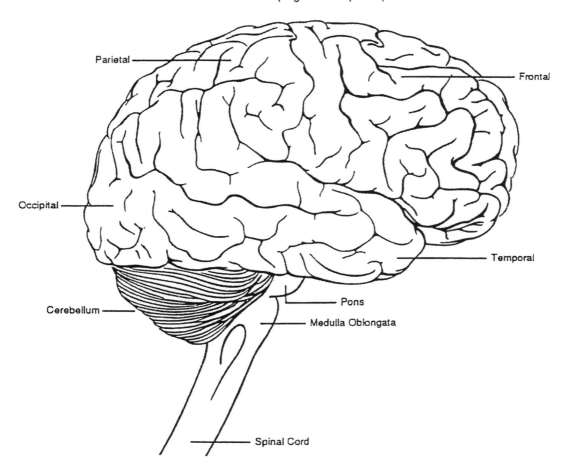

Parietal

Frontal

Occipital

Temporal

Pons

Cerebellum

Medulla Oblongata

Spinal Cord

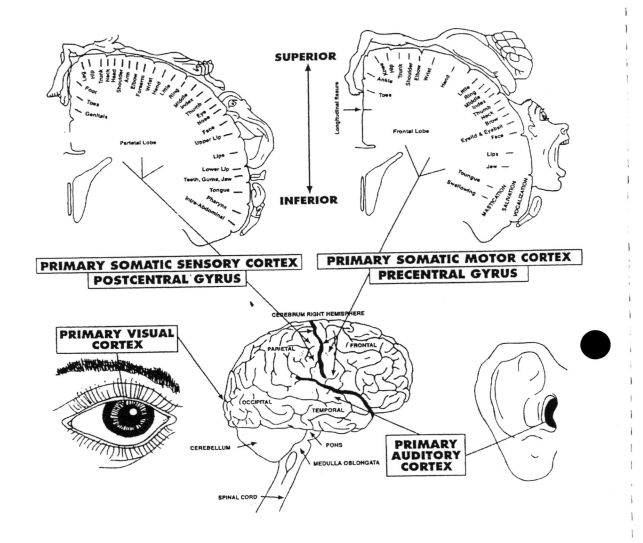

SUPERIOR

INFERIOR

Leg
Hip
Trunk
Neck
Head
Shoulder
Arm
Elbow
Forearm
Wrist
Hand
Little
Ring
Middle
Index
Thumb
Eye
Nose
Face
Upper Lip
Lips
Lower Lip
Teeth, Gums, Jaw
Tongue
Pharynx
Intra-Abdominal

Foot
Toes
Genitals

Parietal Lobe

Longitudinal fissure

Knee
Hip
Trunk
Shoulder
Elbow
Wrist
Hand
Ankle
Toes
Little
Ring
Middle
Index
Thumb
Neck
Brow
Eyelid & Eyeball
Face
Lips
Jaw
Toungue
Swallowing
MASTICATION
SALIVATION
VOCALIZATION

Frontal Lobe

PRIMARY SOMATIC SENSORY CORTEX
POSTCENTRAL GYRUS

PRIMARY SOMATIC MOTOR CORTEX
PRECENTRAL GYRUS

CEREBRUM RIGHT HEMISPHERE

PARIETAL

FRONTAL

OCCIPITAL

TEMPORAL

PRIMARY VISUAL
CORTEX

CEREBELLUM

PONS

MEDULLA OBLONGATA

PRIMARY AUDITORY
CORTEX

SPINAL CORD

NOTES

DATE : _____

*OUTLINE #*_____

QUESTIONS

1._____
2._____
3._____
4._____
5._____

REVIEW QUESTIONS

I.   Matching

A.   Frontal lobe cerebral cortex
B.   Temporal lobe cerebral cortex
C.   Parietal lobe cerebral cortex
D.   Occipital lobe cerebral cortex
E.   Cerebellar cortex

_C_ 1.   feeling the pain of stubbing your toe
_A_ 2.   planning, anticipating what needs to be done, sorting trivial from important, controlling motor related behaviors
_B_ 3.   stores information about the melody of a tune you are going to whistle
_D_ 4.   differentiates between red and purple, blue and green, light and dark
_A_ 5.   recognition of the difference between a C# played on a piano and a C# played on a trumpet

II.   Multiple choice

_E_ 6.   Which of the following would you expect to observe in a patient with a tumor in the cerebellum?
A.   loss of vision *occipital*
B.   loss of hearing *temporal or parietal*
C.   inability to perceive pain
D.   inability to initiate voluntary movements
E.   inability to execute smooth, steady movements

_D_ 7.   Blindfolded, a person is unable to identify the shape of a wooden cube placed in the left hand although the person can sense when the object was placed in the hand. The most probable location of a lesion is the:
A.   right temporal association cortex
B.   Broca's area left cortex
C.   right prefrontal cortex
D.   right parietal association cortex
E.   left primary somatic sensory cortex

_B_ 8.   That portion of the brainstem concerned with visual reflexes such as the pupillary reflex is the:
A.   pons
B.   mesencephalon
C.   medulla oblongata
D.   thalamus
E.   hypothalamus

_E_ 9.　This division of the brain controls visual reflexes such as accommodation of the lens. It is the:

A.　hypothalamus
B.　medulla oblongata
C.　pons
D.　thalamus
E.　mesencephalon

notes

_B_ 10.　Localization and differentiation of thermal vs. mechanical (pressure) cutaneous stimuli involves the _____ cortex of the cerebrum

A.　frontal
B.　parietal
C.　temporal
D.　occipital
E.　occipito-temporal

Pg 130 outline notes

_A_ 11.　Which of the following is (are) dominated by the left cerebral hemisphere?

A.　solving math problems.
B.　artistic creativity —
C.　intuitive thought —
D.　understanding jokes
E.　two of the above

psych book

_B_ 12.　A blindfolded patient can feel or sense the weight of an object as well as its texture (rough, smooth, etc.) when placed in his/her right hand but cannot identify its shape (cube, sphere, disc, etc.) based on how it feels. A probable area of brain damage is the:

A.　left frontal association cortex
B.　left parietal association cortex
C.　right temporal association cortex
D.　right occipital association cortex
E.　left hypothalamus

R Hand

13.　Which of the following is most likely to be associated with a lesion (area of damage) of the frontal cortex?

A.　partial anesthesia
B.　flaccid paralysis
C.　altered mood
D.　spastic paralysis
E.　partial deafness

_E_ 14.　Primary control of heart rate and blood pressure is a function of the:

A.　thalamus
B.　mesecephalon
C.　tenth cranial nerve
D.　medulla oblongata
E.　two of the preceding

Notes

133

_E_ 15.   You reach to pick up an object on a table. The command to pick up was issued by the _____ and the part of the brain ensuring coordination of appropriate muscles and overall smoothness of the movement is the _____.
          A.   medulla, pons
          B.   cerebellum, cerebrum
          C.   thalamus, hypothalamus
          D.   cerebrum, mesencephalon
          E.   cerebrum, cerebellum

_D_ 16.   Which of the following is associated with damage to the cerebellum?
          A.   spastic paralysis
          B.   flaccid paralysis
          C.   past-pointing
          D.   inability to sense body position
          E.   loss of myotatic reflexes

_B_ 17.   A blindfolded person can sense the weight of an object as well as its texture when placed in his/her left hand but cannot identify its shape. A probable area of brain damage is the:
          A.   left frontal association cortex
          B.   right parietal association cortex
          C.   left temporal association cortex
          D.   right occipital association cortex
          E.   left hypothalamus

_E_ 18. _D_ Which of the following is (are) not dominated by the right cerebral hemisphere?
          A.   artistic creativity
          B.   intuitive thought
          C.   subtle meanings of language
          D.   logical thought as in science
          E.   two or more of the preceding

_A_ 19.   Which of the following (is) are dominated by the right cerebral hemisphere?
          A.   artistic creativity
          B.   solving math problems
          C.   motor control of speech
          D.   logical thought
          E.   two of the preceding

_C_ 20.   Broca's area of the brain:
          A.   controls hearing
          B.   is usually in the right cerebrum
          C.   controls speech
          D.   involves taste
          E.   two of the preceding

134

_C_ 21. The primary somatic sensory cortex:
   A.   receives information from the organ of Corti
   B.   is located in the occipital lobe of the cerebrum
   C.   perceives pain
   D.   is located in the frontal lobe of the cerebrum *motor*
   E.   stores information regarding somatic sensations such as what it feels like to touch a hot stove *Temporal*

_B_ 22. The structure of the cerebrum that functionally and anatomically connects the two hemispheres is called the:
   A.   thalamus
   B.   corpus callosum
   C.   hypothalamus
   D.   cerebellum
   E.   superior colliculi

_B_ 23. Before sensory information reaches the cerebral cortex, it is processed and integrated by the:
   A.   cerebellum
   B.   thalamus
   C.   hypothalamus
   D.   brainstem
   E.   alpha motor neurons

_E_ 24. Cutaneous two-point discrimination requires:
   A.   the thalamus
   B.   the 11th cranial nerves
   C.   the hypothalamus
   D.   the parietal cerebral cortex
   E.   two of the preceding

_E_ 25. In the majority of people, which of the following is not hemisphere specific (ie, involving one hemisphere more than the other)?
   A.   artistic creativity
   B.   intuitive thought
   C.   subtle meanings of language
   D.   logical thought processes, as in science
   E.   localization and identification of somatic sensory stimuli

_B_ 26. The area of the brain responsible for the comprehension of language is:
   A.   Broca's area
   B.   Wernicke's area
   C.   prefrontal cortex
   D.   superior colliculus
   E.   medulla oblongata

OUTLINE 10

LEARNING OBJECTIVES (√)

After reading the assigned pages in the textbook, and/or reading and performing related laboratory exercises, and listening to lecture and/or laboratory presentations, you should be able to:

☐1. Diagram the general organization of the autonomic nervous system (ANS) including the following components: CNS nuclei and centers, preganglionic neurons, ganglia, postganglionic neurons, and effectors.

☐2. Identify the neurotransmitters used centrally and peripherally, including preganglionic and postganglionic neurotransmitters.

☐3. Define the sympathetic division of the ANS in terms of CNS outflow, location of ganglia, length of pre and postganglionic fibers, and the locations and types of adrenergic and cholinergic receptors.

☐4. List several target organs and explain the effects on each organ of sympathetic stimulation.

☐5. Define the parasympathetic division of the ANS in terms of CNS outflow, location of ganglia, length of pre and postganglionic fibers, and the locations and types of cholinergic receptors.

☐6. List several target organs and explain the effects on each organ of parasympathetic stimulation.

☐7. Explain why neither division, sympathetic nor parasympathetic, can be thought of as being exclusively excitatory or inhibitory.

☐8. Define autonomic divergence and explain how and why the sympathetic and parasympathetic divisions differ in their degree of divergence.

☐9. Explain the relation between presynaptic inhibition and autonomic tone, using the heart as an example of a target organ displaying the effects of autonomic tone.

☐10. Give an example of a blocker for each of the following autonomic receptors and explain the expected results when the blocker is used: nicotinic, muscarinic, alpha 1, alpha 2, beta 1, and beta 2.

NOTES

DATE : _____   *OUTLINE #_____*

QUESTIONS

1._____
2._____
3._____
4._____
5._____

OUTLINE 10

LECTURER:_____   NOTES_____   DATE:_____

I.   **AUTONOMIC ORGANIZATION** (Diagrams attached)
     A.   CNS Nuclei and Centers

     B.   Autonomic Outflow: general organization
          1.   Preganglionic fibers
          2.   Postganglionic fibers
          3.   Autonomic ganglia
          4.   Neurotransmitters

II. **SYMPATHETIC DIVISION**
     A.   Distribution
          1.   Thoracolumbar outflow
          2.   Ganglia: paravertebral, peripheral

     B.   Major Functions: maintenance of homeostasis in acute stress

III. **PARASYMPATHETIC DIVISION**
     A.   Distribution
          1.   Craniosacral outflow
          2.   Ganglia: visceral, peripheral

     B.   Major Functions: daily routine maintenance of homeostasis

NOTES

QUESTIONS

1._____
2._____
3._____
4._____
5._____

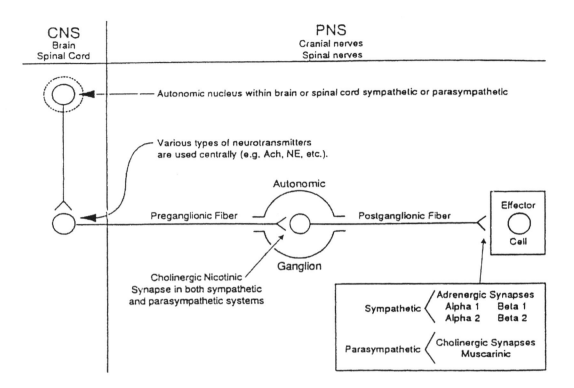

General organization of the Autonomic
Nervous System (ANS); sympathetic
and parasympathetic divisions.

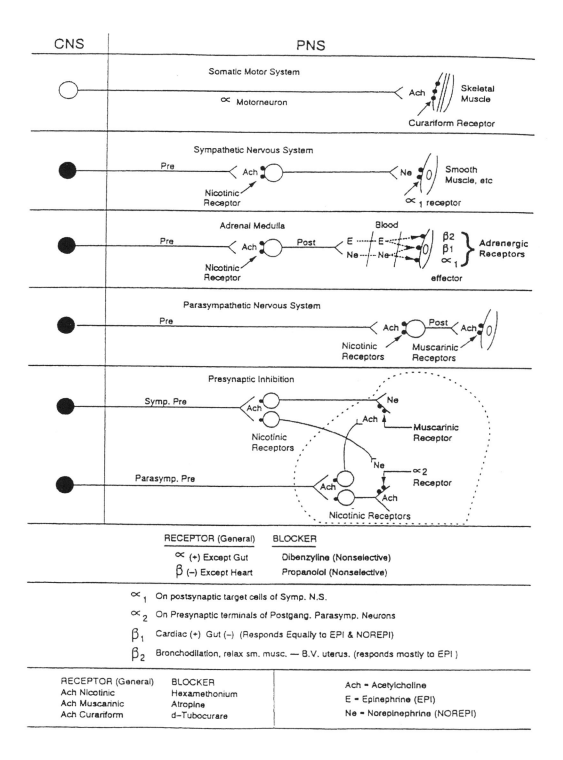

CNS	PNS

Somatic Motor System

∝ Motorneuron

Ach → Skeletal Muscle

Curariform Receptor

Sympathetic Nervous System

Pre — Ach — Nicotinic Receptor

Ne → Smooth Muscle, etc

∝₁ receptor

Adrenal Medulla

Pre — Ach — Nicotinic Receptor — Post

Blood

E ---- E ----
Ne ---- Ne ----

β2
β1
∝₁

Adrenergic Receptors

effector

Parasympathetic Nervous System

Pre — Ach — Nicotinic Receptors — Post — Ach — Muscarinic Receptors

Presynaptic Inhibition

Symp. Pre — Ach — Nicotinic Receptors

Ne
Ach → Muscarinic Receptor
Ne → ∝2 Receptor

Parasymp. Pre — Ach — Nicotinic Receptors — Ach

RECEPTOR (General)	BLOCKER
∝ (+) Except Gut	Dibenzyline (Nonselective)
β (−) Except Heart	Propanolol (Nonselective)

∝₁   On postsynaptic target cells of Symp. N.S.

∝₂   On Presynaptic terminals of Postgang. Parasymp. Neurons

β_1 Cardiac (+) Gut (−) (Responds Equally to EPI & NOREPI)

β_2 Bronchodilation, relax sm. musc. — B.V. uterus. (responds mostly to EPI.)

RECEPTOR (General)	BLOCKER	
Ach Nicotinic	Hexamethonium	Ach = Acetylcholine
Ach Muscarinic	Atropine	E = Epinephrine (EPI)
Ach Curariform	d−Tubocurare	Ne = Norepinephrine (NOREPI)

IV.   AUTONOMIC DIVERGENCE
Diagrams: divergent neuron pathways
A.   Sympathetic Fibers: high degree of divergence

B.   Parasympathetic Fibers: low degree of divergence

V.   AUTONOMIC TONE
A.   Definition and significance

B.   Presynaptic Inhibition: postganglionic neurons
Diagram: autonomic control of heart rate

VI.   AUTONOMIC RECEPTORS AND BLOCKERS
A.   Cholinergic receptors
   1.   Nicotinic: sympathetic and parasympathetic

   2.   Muscarinic: parasympathetic

   3.   Cholinergic blockers

B.   Adrenergic
   1.   Alpha 1, alpha 2

   2.   Beta 1, beta 2

   3.   Adrenergic blockers

NOTES

DATE : _____

OUTLINE #_____

QUESTIONS

1._____
2._____
3._____
4._____
5._____

REVIEW QUESTIONS

I. Matching

A. Alpha 1 receptor
B. Alpha 2 receptor
C. Beta 1 receptor
D. Beta 2 receptor
E. Nicotinic receptor

_____ 1. both sympathetic and parasympathetic divisions of the ANS are affected when this receptor is blocked
_____ 2. engagement of this receptor causes heart rate to increase
_____ 3. engagement of this receptor relaxes smooth muscle in the airways
_____ 4. engagement of this receptor contracts smooth muscle in arterioles of the gut
_____ 5. Found on the postsynaptic neuron terminal bouton, it is involved in presynaptic inhibition

II. Multiple choice

_____ 6. Which type of receptor is found on target cells of preganglionic parasympathetic nerve fibers?
  A. beta
  B. alpha
  C. muscarinic
  D. curariform
  E. nicotinic

_____ 7. Autonomic alpha-one adrenergic receptors are:
  A. blocked by norepinephrine
  B. sympathetic receptors that mediate contraction of vascular smooth muscle
  C. stimulated by alpha-two receptors
  D. excitatory everywhere except the heart
  E. extinct

_____ 8. One of the following statements applies to the sympathetic division of the ANS. Which one?
  A. The ganglia are usually within walls of viscera.
  B. The receptor for postganglionic transmitter is muscarinic.
  C. None of the receptors for transmitters are nicotinic.
  D. Atropine blocks the receptor for postganglionic neurotransmitter.
  E. Activation of the beta receptor usually causes smooth muscle to relax.

9. Which of the following nerves contain preganglionic sympathetic fibers:
A. tenth thoracic spinal nerve ?
B. trigeminal nerve
C. eighth cervical nerve
D. facial nerve
E. optic nerve

10. Which type of receptor is found on target cells of postganglionic parasympathetic nerve fibers?
A. beta receptor
B. alpha receptor
C. muscarinic receptor
D. nicotinic receptor
E. curariform receptor

11. Which of the following reflects increased sympathetic activity?
A. vasodilation in skeletal muscle
B. decreased airway diameter
C. decreased heart rate
D. pupillary constriction
E. two of the preceding

12. Alpha receptors:
A. are cholinergic
B. are muscarinic
C. are nicotinic
D. are adrenergic
E. can be blocked by curare

13. Which of the following describes autonomic tone?
A. The sympathetics excite; the parasympathetics inhibit.
B. Most viscera have sympathetic and parasympathetic innervation.
C. The two divisions of the ANS usually oppose each other and are always active, although their levels of activity vary.
D. Alpha and beta receptors have opposite effects.
E. Two of the preceding.

14. Nicotinic receptors:
A. are adrenergic receptors
B. can be found on both sympathetic and parasympathetic neurons
C. respond to muscarine
D. are the same as curariform receptors
E. are designated as alpha or beta receptors

15. Blocking nicotinic autonomic receptors would result in:
A. decreasing sympathetic but not parasympathetic activity
B. decreasing parasympathetic but not sympathetic activity
C. increasing sympathetic and decreasing parasympathetic activity
D. increasing parasympathetic and decreasing sympathetic activity
E. decreasing both parasympathetic and sympathetic activity

_B_ 16.   Autonomic alpha-one adrenergic receptors are:

  A.   blocked by nicotine
  B.   sympathetic receptors that mediate contraction of vascular smooth muscle
  C.   stimulated by alpha-two receptors
  D.   excitatory everywhere except the heart
  E.   two of the preceding

_A_ 17.   Which of the following nerves contains preganglionic sympathetic fibers?

  A.   tenth thoracic spinal nerve
  B.   trigeminal nerve
  C.   eighth cervical spinal nerve
  D.   facial nerve
  E.   two of the preceding

_C_ 18.   In the autonomic nervous system, beta receptors:

  A.   are cholinergic
  B.   can be blocked by curare
  C.   when engaged with transmitter, increase heart rate
  D.   are found only in skeletal muscle
  E.   are muscarinic

_B_ 19.   Which of the following is not part of the fight or flight response?

  A.   increase in the force of heart contraction
  B.   decrease in blood flow to the skeletal muscles
  C.   increase in pulse rate
  D.   conversion of stored food into glucose
  E.   more oxygen is made available by increased breathing

_D_ 20.   When the sympathetic system is activated one sees:

  A.   increased blood flow to the skin
  B.   increased secretion of gastric enzymes
  C.   increased urine production
  D.   increased airway diameter
  E.   increased production of thin watery saliva

_E_ 21.   Daily routine maintenance of visceral activities is the domain of the:

  A.   primary somatic motor cortex
  B.   cerebellum
  C.   premotor cortex in the frontal lobe of the cerebrum
  D.   sympathetic division of the ANS
  E.   parasympathetic division of the ANS   Homeostasis

_C_ 22.   The autonomic receptors on the postganglionic neuron are:

  A.   muscarinic
  B.   adrenergic
  C.   nicotinic
  D.   sensitive to epinephrine
  E.   none of the preceding

146

23. Considering the sympathetic division of the autonomic nervous system, which statement is true:
   A. the preganglionic transmitter is norepinephrine
   B. the postganglionic transmitter is acetylcholine except in skeletal muscle
   C. the preganglionic neurons are located in the lumbar and sacral spinal cord segments
   D. most postganglionic neurons are very short and located within the walls of viscera
   E. none of the preceding statements are true

24. Which of the following responses is not considered to be sympathetic?
   A. widening of the pupil
   B. accommodation for far vision
   C. inhibition of gastrointestinal functions
   D. increased depth and rate of respirations
   E. conversion of glucose to glycogen in the liver

25. Which one of the following characteristics is *not* representative of the sympathetic division of the ANS?
   A. upon activation, usually targets single organs to produce specific responses
   B. always has at least two synapses after leaving the CNS
   C. releases norepinephrine at most of its postganglionic synapses
   D. activates alpha and beta receptors
   E. contains paravertebral and peripheral ganglia

26. Parasympathetic stimulation:
   A. increases gastrointestinal motility
   B. increases heart rate
   C. increases airway diameter
   D. increases pupil diameter
   E. increases production of saliva that is very thick causing a dry cotton-mouth feeling

OUTLINE 11

LEARNING OBJECTIVES (√)

After reading the assigned pages in the textbook, and/or reading and performing related laboratory exercises, and listening to lecture and/or laboratory presentations, you should be able to:

☐1. Define and give examples of interoceptors and exteroceptors.

☐2. Identify the following integumentary receptors and their adequate stimuli: Pacinian corpuscle, Meissner's corpuscle, Krause's receptor, Ruffini's receptor, Merkel's discs, and the hair ending plexus.

☐3. Describe the following attributes of sensation: modality, quality, quantity, and projection or localization.

☐4. Define and give an example of the law of specific nerve energies.

☐5. Define and give an example of the law of adequate stimulus.

☐6. Use the Pacinian corpuscle to explain the ionic basis of the generator potential.

☐7. Use a hair cell of the inner ear to explain how a receptor potential can generate an impulse in a sensory neuron.

☐8. Outline olfactory receptor structure and function, and olfactory pathways from the olfactory nerve to the cerebral cortex.

☐9. Briefly describe the classification of odor.

☐10. Outline taste receptor structure and function, and gustatory pathways from taste buds to the gustatory cortex of the cerebrum.

☐11. Describe primary taste areas of the tongue and pharynx.

☐12. Define the following parts of the outer ear and briefly describe their function: pinna, external auditory canal, hair follicles , ceruminous glands.

☐13. Define the following parts of the middle ear and briefly describe their function: tympanic membrane, malleus, incus, stapes, Eustachian tube, and muscles of the ossicles (e.g. stapedius).

☐14. Explain in functional terms the difference between the cochlear division and the vestibular division of the inner ear.

☐15. Diagram the cochlea and include the following: scala vestibuli, scala tympani, scala media (cochlear duct), basilar membrane, vestibular membrane, tectorial membrane, hair cells, supporting cells, perilymph, endolymph, and spiral ganglion.

☐16. Explain the mechanism of hearing, beginning with sound entering the external auditory canal and ending with nerve impulses being transmitted along the auditory nerve.

☐17. Explain how sound frequency and sound intensity are discriminated by the organ of Corti.

☐18. Diagram the peripheral and central auditory pathways, showing that both ears are represented in each temporal lobe.

NOTES

DATE : _____   *OUTLINE #*_____

QUESTIONS

1._____
2._____
3._____
4._____
5._____

OUTLINE 11

I. **GENERAL AND SOMATIC SENSORY PHYSIOLOGY**
    A. Sensory Receptors
        1. Classification: exteroceptors and interoceptors

        2. Integumentary receptors: general structure & adequate stimulus
            a. Pacinian corpuscle
            b. Meissner's corpuscle
            c. Krause's receptor
            d. Ruffini's receptor
            e. Merkel's disc
            f. Hair ending plexus

    B. Attributes of Sensation
        1. Modality
        2. Quality
        3. Quantity
        4. Localization or projection

    C. Law of Specific Nerve Energies

    D. Law of Adequate Stimulus

    E. Generator Potential
       Diagram: Pacinian corpuscle

II. **SPECIAL SENSORY PHYSIOLOLGY**
    A. Receptor Potential
       Diagram: Auditory hair cell

NOTES

DATE : _____   *OUTLINE #_____*

QUESTIONS

1._____
2._____
3._____
4._____
5._____

B.   Olfaction (Smell)
1.   Receptor structure and function

2.   Olfactory pathways
Diagram: olfactory receptors to cerebral cortex

3.   Classification of odor

C.   Gustation (Taste)
1.   Receptor structure and function

2.   Taste pathways
Diagram: taste buds to cerebral cortex

3.   Primary taste areas
Diagram: surface of tongue

D.   Auditory (Hearing)
1.   General structure and function of the ear
   a.   Outer ear: pinna, external auditory canal, hair, cerumen

   b.   Middle ear: tympanic membrane, malleus, incus, stapes, eustachian tube, muscles of the ossicles
   c.   Inner ear: cochlea, vestibular apparatus

2.   Cochlea
   a.   Chambers and membranes
      1.   Scala vestibuli
      2.   Scala media (cochlear duct)
      3.   Scala tympani
      4.   Vestibular membrane
      5.   Tectorial mMembrane
      6.   Basilar mMembrane
   b.   Organ of corti
      1.   Hair cells (inner and outer)
      2.   Supporting cells

NOTES

DATE : _____   OUTLINE #_____

QUESTIONS

1._____
2._____
3._____
4._____
5._____

3.   Cochlear function: detection of sound frequency and intensity
      Diagram: *straightened organ of Corti*
                    low frequency detection at apex
                    high frequency detection at base
                    amplitude of maximum displacement = intensity
                    (loudness)

4.   Auditory pathways: peripheral and central
      Diagram: *both ears represented in each temporal lobe*

NOTES

QUESTIONS

1._____
2._____
3._____
4._____
5._____

REVIEW QUESTIONS

I. Matching

A. Hair cells near base of cochlea
B. Hair cells near apex of cochlea
C. Tectorial membrane
D. Vestibular membrane
E. Basilar membrane

_E_ 1. forms the floor of the scala media (cochlear duct); part of the organ of Corti
_A_ 2. maximally stimulated by 20,000 Hz sound
_C_ 3. hair cell processes are embedded in this structure
_D_ 4. forms the roof of the scala media
_B_ 5. maximally stimulated by 4,000 Hz sound

II. Multiple choice

_C_ 6. Damage to the cranial nerve #1 would most likely interfere with the sense of:
A. balance   D. hearing
B. sight   E. taste
C. smell

_D_ 7. Repeated exposure to the high intensity sounds of hard rock music (noise) has left you with an inability to hear sound frequencies at and near 20,000 Hz. A likely site of damage to the auditory system is the:
A. organ of Corti, apical cochlea
B. semicircular canal, superior
C. saccule
D. organ of Corti, basal cochlea
E. spiral ganglion

_C_ 8. So that the tympanic membrane can vibrate normally, the _____ allows air pressure to become equal on either side.
A. pinna
B. cochlea
C. eustachian tube
D. semicircular canal
E. utricle

_C_ 9. Which of the following is not one of the primary tastes?
A. bitter   D. salt
B. sweet   E. sour
C. pungent

_D_ 10. The ability to taste is dependent on normal sensory functioning of three cranial nerves. They are: *facil glossopharyngeal*
A. 3-4-6
B. 2-4-6
C. 5-7-9
D. 7-9-10 *vagus*
E. 5-7-9

_A_ 11. The eustachian tube functions:
A. to equalize air pressure on both sides of the tympanic membrane
B. to help maintain equilibrium
C. to magnify sounds
D. to protect sounds
E. all of the preceding

_C_ 12. The primary function of the stapes is to:
A. relay vibrations of the malleus to the incus
B. relay vibrations of the tympanic membrane to the malleus
C. relay vibrations of the incus to the vestibule
D. dampen excessive vibration of the tympanic membrane
E. none of the preceding

_C_ 13. Interpreting sound intensity (loudness) involves:
A. the parietal cerebrum
B. the hypothalamus
C. the organ of Corti
D. the utricle
E. two of the preceding

_A_ 14. The constant exposure to loud noise near 20,000 Hz would result in loss of hair cells on the basilar membrane:
A. near the base of the cochlea
B. near the apex of the cochlea
C. at a location midway between the base and the apex
D. at the helicotrema
E. two of the preceding

_C_ 15. The organ of Corti:
A. is responsible for equilibrium
B. is located within the scala tympani
C. is responsible for hearing
D. is located within the scala vestibuli
E. two of the preceding

_E_ 16. Discrimination of sound frequencies and intensities detected by the right ear involves:
A. the right temporal cortex
B. the left temporal cortex
C. the frontal cortex
D. the occipital cortex
E. two of the preceding

_____ 17. A lesion in the right primary auditory cortex could result in:
   A.   complete right ear deafness
   B.   complete left ear deafness
   C.   partial right ear deafness
   D.   partial left ear deafness
   E.   answers (C) and (D)

_____ 18. Conductive hearing loss results from damage to the:
   A.   outer ear
   B.   middle ear
   C.   inner ear
   D.   cerebral cortex
   E.   two of the above

_____ 19. The organ of Corti is:
   A.   located in the scala media
   B.   responsible for hearing
   C.   responsible for equilibrium
   D.   located in the scala tympani
   E.   two of the preceding

_____ 20. In which component of the cochlea do sounds cause waves having the peak
   amplitude of a specific tone (frequency) at a specific location?
   A.   oval window
   B.   organ of Corti
   C.   basilar membrane
   D.   Reissner's membrane
   E.   scala media (cochlear duct)

_____ 21. A lesion in the brainstem affecting auditory pathways could result in a loss
   of hearing classified as:
   A.   sensorineural hearing loss
   B.   conductive hearing loss
   C.   central hearing loss
   D.   peripheral hearing loss
   E.   presbycusis

NOTES

QUESTIONS

1._____
2._____
3._____
4._____
5._____

OUTLINE 12

LEARNING OBJECTIVES (√)

After reading the assigned pages in the textbook, and/or reading and performing related laboratory exercises, and listening to lecture and/or laboratory presentations, you should be able to:

☐1. Define static equilibrium and dynamic equilibrium.

☐2. Explain the roles of the maculae and otoliths of the utricle and saccule in the process of static equilibrium, and in sensing linear acceleration.

☐3. Use the three planes of space to diagram the three semicircular canals and explain how each canal can detect angular acceleration in one of the three planes of space.

☐4. Define post-rotational nystagmus and explain the direction of the fast component of eye movement and the slow component of eye movement when the subject's eyes are open during rotation and when the subject's eyes are closed during rotation.

☐5. Diagram the general structure of the eye. In a longitudinal section, identify the following: retinal, vascular, and supporting layers, chambers, humour, pupillary constrictor, pupillary dilator, lens, suspensory ligament, ciliary muscle, fovea centralis, optic disc, and cornea.

☐6. Define each of the following and describe the shape of the corrective lens (if required): emmetropia, myopia, hypermetropia, and astigmatism.

☐7. Explain what is meant by accommodation of the lens (accommodation reflex) and explain the roles of the ciliary muscle, the suspensory ligament, and the lens itself in the accommodation reflex.

☐8. Define pupillary reflex and explain how the sympathetic and parasympathetic divisions of the ANS can alter pupillary diameter.

☐9. List four major physiological differences between rods and cones of the retina (e.g. visual acuity, color vision, detection of movement, night vision, etc.).

☐10. Diagram the optic pathways from the retina to the occipital cotex, including the following components: optic nerves, optic chiasma, optic tracts, lateral geniculate bodies of the thalamus, geniculocalcarine   tracts, and primary visual cortex of the occipital lobe.

☐11. Explain the loss in temporal and/or nasal visual fields when the following areas of the visual pathway are interrupted: optic nerve, optic tract, optic chiasm, thalamus, primary visual cortex.

NOTES

DATE : _____   *OUTLINE #_____*

QUESTIONS

1._____
2._____
3._____
4._____
5._____

OUTLINE 12

I. **EQUILIBRIUM**
    A.   Static Equilibrium: Utricle & Saccule
           Definition:

           Diagram: head tilted, body stationary

    B.   Dynamic Equilibrium: Semicircular Canals
           Definition:

           Diagram: body rotating around vertical axis

II. **VISION**
    A.   General Eye Structure
           Diagram: midsagittal section through eyeball

        1.   Retinal layer
        2.   Vascular layer
        3.   Supporting layer
        4.   Chambers
        5.   Pupillary constrictor
        6.   Pupillary dilator
        7.   Lens and suspensory ligament
        8.   Ciliary muscle
        9.   Fovea centralis
        10.   Optic disc
        11.   Cornea

NOTES

DATE : _____   *OUTLINE #_____*

QUESTIONS

1._____
2._____
3._____
4._____
5._____

B. Optics of Vision
1. Emmetropia (Normal)

2. Myopia (Near-sightedness)

3. Hypermetropia (Far-sightedness)

4. Astigmatism

C. Visual Reflexes
1. Accommodation *oculomotor nerve*
III +IV +I +VI
Diagram: contraction and relaxation of ciliary muscle

2. Pupillary
Diagram: pupillary constrictor and pupillary dilator muscles

D. Retinal Function
1. Cones

2. Rods

NOTES

DATE : _____   *OUTLINE #_____*

QUESTIONS

1._____
2._____
3._____
4._____
5._____

3.   Neural pathways
     Diagram: transverse section through brain and optic pathways

         a.   Retina, optic nerve

         b.   Optic chiasma

         c.   Optic tract

         d.   Thalamus, geniculo-calcarine tract

NOTES

DATE : _____   *OUTLINE # _____*

QUESTIONS

1._____
2._____
3._____
4._____
5._____

REVIEW QUESTIONS

I. Matching
A.   Emmetropia
B.   Hypermetropia
C.   Myopia
D.   Presbyopia
E.   Astigmatism

_D_ 1.   "old age vision" due to loss of lens elasticity
_C_ 2.   near-sightedness
_B_ 3.   far-sightedness
_A_ 4.   normal visual acuity
_E_ 5.   abnormal curvature of the lens, or cornea

II. Multiple choice

_D_ 6.   Rotation of the body to the left about a vertical axis (like a spinning ice skater) with the subject's eyes open produces a nystagmus that is:
A.   horizontal with the slow component to the left
B.   horizontal with the fast component to the right
C.   vertical with the fast component to the right
D.   horizontal with the fast component to the left
E.   vertical with the slow component to the right

_A_ 7.   Which of the following is not one of the pigments found in cones?
A.   yellow
B.   blue
C.   red
D.   green
E.   none of the above

_A_ 8.   When testing for vestibular function by rotating a subject seated in a swivel chair to the subject's right, in which direction was the fast component of nystagmus?
A.   to the right
B.   to the left
C.   up toward the eyebrows
D.   down toward the mouth
E.   circular, counterclockwise

_C_ 9.   Hypermetropia:
A.   occurs when light is focused in front of the retina
B.   is another name for presbyopia
C.   can be corrected by placing a convex lens in front of the eye
D.   means vision which is extraordinarily good
E.   none of the preceding

_D_10. ✓ The sensory receptors for detecting linear acceleration in a horizontal plane are located in the:
- A. ampulla
- B. scala vestibuli
- C. saccule
- D. utricle
- E. middle ear

_B_11. When the cilia of the hair cells in the right lateral semicircular canal are bent toward the kinocilium, the hair cell membrane:
- A. hyperpolarizes
- B. depolarizes
- C. first hyperpolarizes, then depolarizes
- D. moves closer to a threshold potential
- E. remains at resting membrane potential

_A_12. Linear acceleration in a vertical plane is detected by receptors in the:
- A. saccule
- B. cochlea
- C. utricle
- D. semicircular canals
- E. organ of Corti

_D_13. Contraction of the ciliary muscle causes:
- A. the eyeball to rotate laterally
- B. the pupil to become smaller
- C. the lens to become thinner (flatten)
- D. the lens to become thicker (bulge)
- E. the near-point to recede

_B_14. The right optic tract has been compressed by a tumor, blocking transmission in the optic fibers. The loss of vision:
- A. would exclusively involve the right eye
- B. would involve the visual fields of both eyes
- C. would involve the visual cortex of both right and left cerebral hemispheres
- D. would result in complete blindness
- E. would be called emmetropia

_D_15. The most likely result of irregular curvature of the cornea is:
- A. myopia
- B. hypermetropia
- C. emmetropia
- D. astigmatism
- E. nyctalopia

C 16. Which of the following is (are) true concerning retinal function?
   A.   Cone vision is best at low levels of light intensity.
   B.   Color vision is mediated by rods.
   C.   Rod vision of low acuity.
   D.   Visual acuity is highest in the peripheral retina.
   E.   None of the preceding are true.

C 17. The trichromatic theory of color vision is based on:
   A.   rods, not cones
   B.   red rods, blue cones
   C.   red, blue, and green cones
   D.   red, white and blue cones
   E.   Dr. Shinobu Ishihara

B 18. The function of the ciliary muscle of the eye is:
   A.   convergence
   B.   to adjust the amount of light entering the eye
   C.   to distinguish between light and dark objects
   D.   to distinguish different colors
   E.   none of the preceding

E 19. The retina contains:
   A.   ganglion cells   D.   horizontal cells
   B.   photoreceptors   E.   all of the above
   C.   bipolar cells

E 20. The sensory information carried by fibers in the optic tract on the right side
   of the brain is derived from:
   A.   the entire right visual field
   B.   the nasal component of the left and right visual fields
   C.   the temporal component of the left and right visual fields
   D.   the nasal component of the right visual field and the temporal
        component of the left visual field
   E.   the nasal component of the left visual field and the temporal
        component of the right visual field

A 21. The visual defect corrected by a concave lens that occurs because the image is
   focused in front of the retina is:
   A.   myopia
   B.   hypermetropia
   C.   astigmatism
   D.   emmetropia
   E.   cyclopia

_____D___22.   As an object in your visual field is moved closer to your eyes, the visual
image of the object remains clear because:
A.   the ciliary muscle relaxes
B.   your eyes diverge
C.   you have passed the near point
D.   the ciliary muscle contracts
E.   the pupils constrict

____A___23.   Cones are:
A.   receptors for color vision
B.   found only in the optic disc
C.   receptors for night vision
D.   found primarily in the periphery of the retina
E.   two of the preceding

____D___24.   Angular acceleration, as in somersaulting, is detected by receptors located in
the:
A.   scala tympani
B.   scala media
C.   organ of Corti
D.   semicircular canals
E.   cochlea

____E___25.   Damage to the left lateral geniculate nucleus could result in:
A.   partial loss of hearing from the right ear
B.   partial loss of sight from the right eye
C.   partial loss of sight from the left eye
D.   a loss of somatic sensation on the left side of the body
E.   B and C

OUTLINE 13

LEARNING OBJECTIVES (√)

After reading the assigned pages in the textbook, and/or reading and performing related laboratory exercises, and listening to lecture and/or laboratory presentations, you should be able to:

☐1. Define whole blood, blood plasma, and serum.

☐2. Give normal adult values for each sex for blood volume, specific gravity of whole blood, viscosity of whole blood, viscosity of plasma, and average osmotic pressure (whole blood).

☐3. Describe the physical characteristics of a normal erythrocyte and give normal adult values for the following: RBC count, erythrocyte sedimentation rate, hematocrit ratio, hemoglobin content (whole blood), and oxygen-carrying capacity (whole blood).

☐4. Compute the following red cell indices and give normal values: mean corpuscular volume (MCV), mean corpuscular hemoglobin (MCH), and mean corpuscular hemoglobin concentration (MCHC).

☐5. Define anemia, characterize its symptoms, and explain the classification of anemia based on cell size, shape, and color.

☐6. Define polycythemia, list several types of polycythemia and one cause of each.

☐7. Explain the role of erythropoietin in red cell formation.

☐8. Diagram the hemoglobin molecule and explain its fate when red cells are destroyed. Include the metabolism of iron.

☐9. Describe the appearance of the following white blood cells in a Wright-stain smear: neutrophil, eosinophil, basophil, lymphocyte, and monocyte.

☐10. List one unique function of each of the cells listed in objective #9.

☐11. Define the following leukocyte activities: phagocytosis, chemotaxis, and secretion.

☐12. Make simple diagrams to explain the processes of an antibody-mediated immune response and a cell-mediated immune response.

☐13. Give normals for a combined white blood cell count and a differential white blood cell count.

☐14. Define leukocytosis, leukopenia, leukemia, and give examples of white blood cell counts associated with each.

☐15. List four mechanisms of hemostasis and explain how each contributes to minimizing blood loss.

☐16. Diagram the three stages of blood coagulation, and explain the mechanism of action of two anticoagulants and two thrombolytics.

☐17. Define major and minor agglutination.

☐18. List whole blood donor types for each recipient type in the ABO-Rh groups.

NOTES

QUESTIONS

1._____
2._____
3._____
4._____
5._____

NORMAL ADULT BLOOD

MEASUREMENT	VALUES
Blood Volume (6-8% total body weight)	4.5 – 5.5 liters (female)
	5.0 – 6.0 liters (male)
Specific Gravity (whole blood)	1.050 – 1.060
Viscosity (whole blood)	$3.5 – 5.5 \times H_2O$
Viscosity (plasma)	$1.9 – 2.6 \times H_2O$
Erythrocyte Sedimentation Rate (ESR)	2.0 – 20.0 mm/hr (female)
	2.0 – 10.0 mm/ hr (male)
Hermatocrit	$42\% \pm 5\%$ (female)
	$47\% \pm 5\%$ (male)
Average Osmotic Pressure (whole blood)	7.3 atm. (5550 torr)
Hemoglobin Content	12 – 16 gms/dL (female)
	14 – 18 gms /dL (male)
O_2 Carrying Capacity	$16 – 25$ ml O_2/dL
Erythrocyte Count (RBC)	4.2 – 5.4 million/μL (female)
	4.6 – 6.2 million/μL (male)

Red Cell Indices

Mean Corpuscular Volume (MCV)	$82 – 92 \ \mu m^3$
Mean Corpuscular Hemoglobin (MCH)	27 – 31 $\mu\mu$g or pg
Mean Corpuscular Hemoglobin Concentration (MCHC)	32 – 36%
White Blood Cell Count (WBC)	7,000 – 10,000 cells/μL
Differential WBC (%):	

Neutrophils	50 – 70%
Eosinophils	1 – 5%
Basophils	0 – 1%
Lymphocytes	20 – 40%
Monocytes	1 – 6%

Normal Blood slightly alkaline 7.35 - 7.45

platelets = thrombocytes

plasma - extracellular Fluid

Blood Fractions = RBC, WBC, platelets, plasma, and protien fractions of WBC (albumin, immunoglobin, and clotting factors) Pg 522

Polycythemia = too many RBC's - decrease in RBC sedim.

Plasma - coagulates - 93% water

serem - doesn't coagulate

70 kg male → 28L intercellular Fluid - 10.5 L interstitial fluid
       3.5 L Blood plasma

QUESTIONS

1._____
2._____
3._____
4._____
5._____

OUTLINE 13

LECTURER:_____   NOTES_____   DATE:_____

I.   **BLOOD**
  A.   Definition: liquid connective tissue

  B.   Composition: variable

II.   **BLOOD CELLS**
  A.   Erythrocytes (Red Blood Cells)
    1.   Physical characteristics
    Diagram: diameter, thickness, shape Biconcave
    8.5um x 2.5 um

    2.   Count and hematocrit: adult male, female

    3.   Hemoglobin and iron: adult male, female
    Diagram: iron metabolism and fate of hemoglobin

    Hemoglobin=
    1 globin molecule +
    4 Heme groups attached
    can carry 4 oxygen molecules

    Saturated in Lungs
    becomes oxyhemoglobin
    bright red!!!

    4.   Oxygen-carrying capacity of whole blood
    calculation: Hb content (gm/dl) times 1.34 ml O_2/gm
    15g/dL x 1.34 mL O2/g = 20.1 mL O2/dL

    5.   Anemia
      a.   Definition: reduction in O_2 carrying capacity

      b.   Symptoms and types

NOTES

DATE : _____   OUTLINE #_____

QUESTIONS

1._____
2._____
3._____
4._____
5._____

FATE OF HEMOGLOBIN

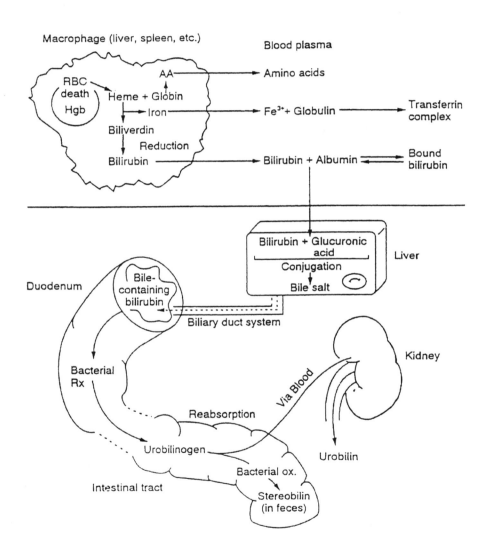

NOTES

QUESTIONS

1._____
2._____
3._____
4._____
5._____

6. Polycythemia
    a. Definition

High RBC count

    b. Polycythemia vera

    c. Physiological polycythemia

B. Leukocytes (Combined Count = 7000-10,000 cells/μl)

come from: (PHSC) pluripotential hemopoietic stem cells

  1. Major types and counts: granulocytes, agranulocytes
    a. Neutrophil count: *50-70%*
      Functions:

    b. Eosinophil count: *1-5%*
      Functions:

    c. Basophil count: *0-1%*
      Functions:

    d. Monocyte count: *1-6%*
      Functions:

    e. Lymphocyte count *20-40%*
      Functions:

  2. General leukocyte functions
    a. Diapedesis

    b. Chemotaxis

    c. Phagocytosis

    d. Secretion

NOTES

DATE : _____

*OUTLINE #*_____

QUESTIONS

1._____
2._____
3._____
4._____
5._____

3. Cell-mediated immunity
   Diagram: CD4+ cells and CD8+ cells

4. Antibody-mediated immunity
   Diagram: plasma cells and antibody production

5. Leukocytosis

6. Leukopenia

III. **PLASMA**
   A Plasma vs. Serum - coagulating factors vs none
   B. Physical Characteristics transparent light yellow
   C. Chemical Composition -93% water 7% solutes

IV. **HEMOSTASIS**
   A. Definition: blood loss prevention

   B. Processes of Hemostasis

Plasma Protiens: 60% Albumin
                   80% of colliod Pressure (osmotic)

Reduction due to diseases → Kwashiorkor (nutrional), Liver or
    Kidney disease moves extra fluid out of blood
    Resulting in Peripheral Edema
                   40% Globulins

4% a_1-globulin, 8% a_2-globulin, 7% B_1-globulin
4% B_2-globulin, 17% y-globulin

a_1-globulin = glycoprotiens (protien + carbohydrates)
                 lipoprotiens (protien + lipid)
          High Density lipoprotiens (HDL)
(cortisol - binding (transcortin) Vit. B12 Binding (transcobalamino)
(thyroxine-binding-globulins - all transport respective substrates

$a_2$globulins = haptoglobin (Combines with free hemoglobin)
           ceruloplasmin (copper containing oxidase enzyme)
           prothrombin (coagulation)
           Erythropoietin (hormone erythrocyte production)
           angiotensinogen (reg. BP, Body fluid + electrolyte balance)

B Globulins $_{1+2}$ = apolipoprotiens (carriers for lipids)
           B. lipoprotien (LDL) low Density lipoprotien - sticks to art. walls
     phospholipids, glycerides, lipid soluble vitaminsADEK
           Transferrin - transports copper + iron
y-globulins - Immunoglobulins (Ig) antibodies
more than 99% are A, G, M. D + E are rare in plasma

Fibrinogen converts to Fibrin for blood clotting

QUESTIONS

1._____
2._____
3._____
4._____
5._____

183

C.   Coagulation: definition and general mechanism

1.   Stage I *Formation of prothrombin converting factor*

2.   Stage II *Conversion of prothrombin to thrombin*

3.   Stage III *Conversion of fibrinogen to fibren*

D.   Anti-Coagulants and Clot Busters

V.   BLOOD GROUPS

A.   Antigen vs. Antibody
1.   Antigen definition *chemical that stimulates B.cells to produce antibodies*
2.   Antibody definition *Protein produced by B cells in response to nonself antigen*
3.   An antigen-antibody reaction: agglutination
Clamping of Blood cells in response to a reaction between antibody + antigen

B.   ABO - Rh Blood Groups

In Plasma antibodies

TYPE	Ag (RBC) *Antigen*	Ab (Plasma)
A	A	B
B	B	A
AB	A & B	No a, no b
O	No A, no B	a & b
Rh+	D	No d
Rh-	No D	D

on RBC

C.   Major vs. Minor Agglutination
Recipients antibodies React to Donor antigen vs. *Donors antibodies react to Recipients antigen*

D.   Eyrthoblastosis Foetalis
Diagram: Rh- mom: Rh + fetus #1, #2

184

NOTES

DATE : _____   *OUTLINE #_____*

QUESTIONS

1._____
2._____
3._____
4._____
5._____

ABO GROUP

ANTIBODIES < a = ANTI-A
            b = ANTI-B

AGGLUTINATION OF RBC

Found in serum called ANTI-A serum (serum from Type B blood). Also found in serum from Type O blood.

Found in serum called ANTI-B serum (serum from Type A blood). Also found in serum from Type O blood.

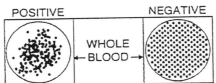

POSITIVE | NEGATIVE

WHOLE BLOOD →

ADD
ANTI A SERUM | ANTI B SERUM

WHOLE BLOOD TYPE A

WHOLE BLOOD TYPE B

WHOLE BLOOD TYPE AB

WHOLE BLOOD TYPE O

RBC TYPES

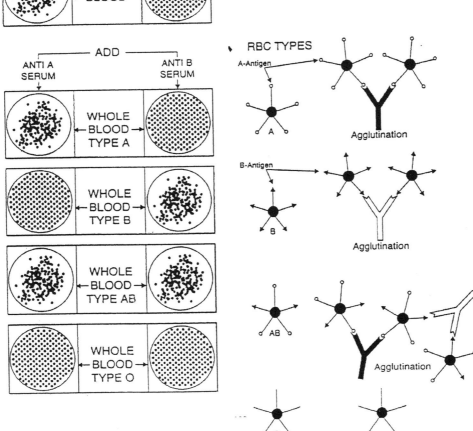

A-Antigen
A
Agglutination

B-Antigen
B
Agglutination

AB
Agglutination

O
No Agglutination

186

NOTES

QUESTIONS

1._____
2._____
3._____
4._____
5._____

REVIEW QUESTIONS

I. Matching

A. Hemoglobin content, adult male
B. RBC count, adult female
C. Lymphocyte count, adult male
D. Hematocrit, adult female
E. Mean Corpuscular Volume (MCV), adult male

_E_ 1. 90 μm³
_B_ 2. 4.8 million/μl
_D_ 3. 40%
_C_ 4. 70%
_A_ 5. 15 g/dl

II. Multiple choice

_D_ 6. In a 20 ml sample of whole blood, packed red cell volume is 9 ml. The hematocrit is therefore:
   A. 1190
   B. 4.5 ml
   C. 11 ml
   D. 45%
   E. 45 ml

_B_ 7. The effectiveness of both antibody-mediated immune responses and cell-mediated immune responses to the presence of an antigen is most dependent on:
   A. eosinophils
   B. CD4$^+$lymphocytes (T$_4$ cells)
   C. CD8$^+$ lymphocytes (T$_8$ cells)
   D. plasma cells
   E. basophils

_C_ 8. During blood coagulation, a soluble plasma protein produced by the liver is converted to an insoluble plasma protein. The conversion is the main event of:
   A. Stage I
   B. Stage II
   C. Stage III
   D. fibrinolysis
   E. kwashiorkor

C 9. Which of the following is a renal hormone that stimulates hematopoiesis?
   A. aldosterone
   B. enterogastrone
   C. erythropoietin
   D. prothrombin
   E. estrogen

B 10. The appropriate blood volume of a 70 kgm adult is:
   A. 3 liters
   B. 5 liters
   C. 9 liters
   D. 12 liters
   E. 15 liters

A 11. Which of the following values for platelet count would be indicative of thrombocytopenia? *Lack of enough Platlets*
   A. 20,000 per μl
   B. 200,000 per μl
   C. 2 million/μl
   D. 400,000 per μl
   E. Two of the preceding

D 12. If hemoglobin content is 13 grams per deciliter and the combining power of hemoglobin for oxygen is 1-1/3 ml of oxygen per gram of hemoglobin, then the oxygen-carrying capacity of the blood:
   A. cannot be computed without more data
   B. is approximately 20 vol. percent
   C. is approximately 15 ml of O_2 per liter
   D. is 17-18 ml of O_2 per deciliter of blood
   E. is 17.4 gms/ml

D 13. One of the following is abnormally high. Which one?
   A. WBC: 8000 per microliter
   B. RBC: 5.3×10^6 per microliter
   C. ESR: 6.0 mm/hr
   D. Hematocrit: 60 percent
   E. MCH: 30 $\mu\mu$g (micromicrograms)

14. Which of the following may be infused into an A⁻ recipient without major or minor agglutination occurring?
   A. whole blood type B⁺
   B. serum from type O⁻ blood
   C. serum from type A⁺ blood
   D. whole blood type AB⁻
   E. whole blood type O⁻

189

C ____ 15. Upon death of the erythrocyte, the hemoglobin contained within the cell:
  A.   is completely destroyed and excreted into the intestine
  B.   remains intact and is incorporated into new erythrocytes
  C.   is broken down into heme and globin from which iron and amino acids are extracted and conserved
  D.   remains intact and is excreted into bile
  E.   is released into the plasma and returned to the bone marrow for reuse

Pg 178 out Book

D ____ 16. The fraction of the total blood volume occupied by erythrocytes is:
  A.   elevated in anemia
  B.   normally about 12-16% in females
  C.   depressed in polycythemia
  D.   termed the hematocrit
  E.   higher in females than in males

D ____ 17. Which of the following values are normal for mean corpuscular hemoglobin?
  A.   10-20 $\mu\mu g$
  B.   13-19 $\mu\mu g$
  C.   12-19 $\mu\mu g$
  D.   27-32 $\mu\mu g$
  E.   45-50 $\mu\mu g$

Pg 174 DB

E ____ 18. The following leukocyte plays an important role in the destruction of immune complexes.
  A.   Monocyte
  B.   Neutrophil
  C.   Eosinophil
  D.   Basophil
  E.   Lymphocyte

Pg 544

B ____ 19. If a person's combined WBC were 7,000 cells per microliter, a normal monocyte count would be (in cells per microliter):
  A.   3500
  B.   350
  C.   35
  D.   5
  E. 1×10^3

  7000
  × .05
  350.00

E ____ 20. Which of the following contains antibodies against antigens A, B, and D?
  A.   Serum of whole blood type O^+
  B.   Whole blood type A^-
  C.   Whole blood type B^+
  D.   Plasma of whole blood type AB^+
  E.   None of the preceding

_C_ 21. Which of the following values for cell count (per ul) would indicate polycythemia vera? (all cells produced in marrow)

    A.   neutrophils: 7500
    B.   lymphocytes: 3000
    C.   erythrocytes: 8.5 million
    D.   monocytes: 500
    E.   none of the preceding

_C_ 22. One or more of the following combinations suggests hypochromic anemia. Which one?

    A.   hemoglobin - 12 gms/dl, RBC 4.8 million/μl
    B.   hemoglobin - 18 gms/dl, RBC 5.4 million/μl
    C.   hemoglobin - 9 gms/dl, RBC 4.0 million/μl
    D.   hemoglobin - 14 gms/dl, RBC 5.0 million/μl
    E.   two of the preceding

_C_ 23. Cytotoxic chemical production is the primary function of which cell?

PG 862

    A.   erythrocyte
    B.   eosinophil
    C. T_8 cell (CD8$^+$)
    D. T_4 cell (CD4$^+$)
    E.   B-lymphocyte

OUTLINE 14

LEARNING OBJECTIVES (√)

After reading the assigned pages in the textbook, and/or reading and performing related laboratory exercises, and listening to lecture and/or laboratory presentations, you should be able to:

☐1. Outline four physical requirements of adequate blood circulation.

☐2. Diagram the double circulation (figure 8 pattern) and label the four cardiac chambers, systemic arterial and venous vessels, and pulmonary arterial and venous vessels.

☐3. Quantitatively describe the relation between flow, pressure, and resistance in the systemic circulation and in the pulmonary circulation.

☐4. Diagram blood vessels of the systemic circuit or the pulmonary circuit in the correct sequence of vessel types and describe the relation between the cross-sectional areas of the vascular pathway and the velocity of blood flow.

☐5. Define Poiseuille's law and explain the influence of each factor in Poiseuille's equation on the flow of blood through an artery.

☐6. Explain the difference between the properties of vascular elasticity and vascular distensibility, and explain the contribution of vascular elastic recoil to systemic arterial blood flow.

☐7. Explain why cardiac muscle is a functional syncytium and why a long absolute refractory period is beneficial to the ventricular myocardium.

☐8. Define systole and diastole

☐9. List in the correct sequence the components of the pacemaker system and explain how the pacemaker system determines the timing of atrial systole and diastole and ventricular systole and diastole.

☐10. Begin with diastasis and describe the electrical and mechanical events of one cardiac cycle in their proper time sequence. Include atrial systole and diastole, isovolumetric contraction and relaxation of the ventricles, rapid ejection, rapid filling, and the opening and closing of intracardiac valves.

☐11. Diagram Lead II of the ECG, label the following components and explain what they represent in the cardiac cycle: P-wave, QRS complex, T-wave, PR interval, QT interval, and ST-segment.

☐12. Define myogenic heart and neurogenic heart.

☐13. Give the normal range for an adult resting heart rate and define bradycardia and tachycardia.

☐14. Diagram and explain the functional significance of the carotid sinus reflex and the Bainbridge reflex in controlling heart rate.

NOTES

DATE : _____   *OUTLINE # _____*

QUESTIONS

1. _____
2. _____
3. _____
4. _____
5. _____

OUTLINE 14

I.   **BLOOD CIRCULATION**
   A.   Physical Requirements
      1.   Pumps
      2.   Vasculature
      3.   Blood volume
      4.   Unidirectional flow

   B.   Outline of Human System
      Diagram: appended
      1.   Four-chambered heart
         a.   Atria: receiving chambers
         b.   Ventricles: pumping chambers
      2.   Systemic circuit
         a.   Arterial
         b.   Venous
      3.   Pulmonary
         a.   Arterial
         b.   Venous

   C.   Flow = Pressure/Resistance
      Diagram: appended
      1.   Systemic circulation: pressures

      2.   Pulmonary circulation: pressures

   D.   Flow: Velocity and Cross-Sectional Area
      Diagram: appended

   E.   Poiseuille's Law
      Diagram: single blood vessel
      Formula: flow directly proportional to ΔP and r^4
                 Inversely proportional to L and viscosity

NOTES

QUESTIONS

1._____
2._____
3._____
4._____
5._____

CIRCULATORY PATHWAY

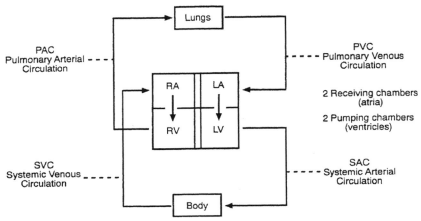

PAC
Pulmonary Arterial
Circulation

PVC
Pulmonary Venous
Circulation

2 Receiving chambers
(atria)

2 Pumping chambers
(ventricles)

SVC
Systemic Venous
Circulation

SAC
Systemic Arterial
Circulation

Lungs

RA   LA

RV   LV

Body

PULMONARY CIRCUIT

RV   LA

SYSTEMIC CIRCUIT

LV   RA

A   AL   C   VL   V

NOTES

DATE : _____　　　　　　　　　　　　　　*OUTLINE #_____*

QUESTIONS

1._____
2._____
3._____
4._____
5._____

F. Flow and Pressure: Vascular Elasticity vs. Vascular Distensibility and their influence on blood flow
Diagram: left ventricle and aorta

II. THE HEART
    A. Functional Gross Anatomy
    Diagram: anterior view of coronal section of heart
        1.   Epicardium, myocardium, endocardium
        2.   Valves
            a.   Semilunar: aortic, pulmonary
            b.   Atrioventricular: bicuspid (mitral), tricuspid
        3.   Papillary muscles, chorda tendinae
        4.   Venous return
            a.   Venae cavae
            b.   Pulmonary veins
        5.   Arterial outflow
            a.   Aorta
            b.   Pulmonary trunk

    B. Histology: Functional Syncytium
    Diagram: interconnected cardiac muscle fibers

    C. Electrical Properties of Ventricular Myocardium: Long Absolute Refractory Period
    Diagram: ventricular action potential and muscle twitch

NOTES

DATE : ________

*OUTLINE #*________

QUESTIONS

1.____________________
2.____________________
3.____________________
4.____________________
5.____________________

D. The Cardiac Cycle
1. Definitions
   a. Systole *-contract*
   b. Diastole *- relax*

2. The Pacemaker system
   Diagram: the cardiac conduction system
   a. SA node
   b. Atrial pathways and atrial muscle
   c. AV node
   d. AV bundle (His)
   e. Rt. & Lt. bundle branches
   f. Purkinje fiber
   g. Ventricular muscle

3. Components of the cardiac cycle, pressure and volume changes
   a. Diastasis
   b. Atrial systole & ventricular diastole
   c. Atrial diastole & ventricular systole

E. Electrocardiography
1. Leads (definition)
   Diagram: lead II electrocardiogram; RA(-)—LL(+)
2. Lead II
   a. P-wave *- systolic depolarization*
   b. QRS complex *- ventricular depolarization*
   c. T-wave *- ventricular Repolarization*
   d. PR interval *- time of impulse between atrium & Ventricle*
   e. QT interval *- ventricular de to re polarization*
   f. ST segment *- refractory ventricular*

F. Neurochemical Control of Heart Rate
1. Normal range: adult heart rate
   a. Bradycardia *- less than 60 Bpm*
   b. Tachycardia *- greater than 100 Bpm*

2. Medullary centers
   a. Cardioaccelerator
   b. Cardioinhibitory

200

NOTES

QUESTIONS

1._____

2._____

3._____

4._____

5._____

[handwritten: Pg 613-14]

[handwritten: monitors variable]

3. Baroreceptor reflexes *[handwritten: cardio vascular sys.]*

   a. Carotid sinus reflex *[handwritten: Reciprocal]*

      Diagram: carotid sinus reflex adjustment of blood
      pressure

[handwritten: center located in medulla]

[handwritten: increase in aterial pressure causes stretch →]

   b. Bainbridge reflex

      Diagram: venous baroreceptor reflex adjustment of
      cardiac output

NOTES

DATE : _____   *OUTLINE #_____*

QUESTIONS

1._____
2._____
3._____
4._____
5._____

EFFECTS OF THE RESTING RESPIRATORY CYCLE ON HEART RATE

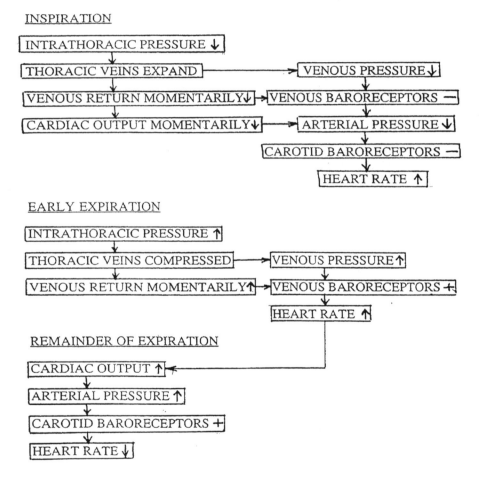

INSPIRATION

INTRATHORACIC PRESSURE ↓

THORACIC VEINS EXPAND → VENOUS PRESSURE ↓

VENOUS RETURN MOMENTARILY↓ → VENOUS BARORECEPTORS —

CARDIAC OUTPUT MOMENTARILY↓ → ARTERIAL PRESSURE ↓

CAROTID BARORECEPTORS —

HEART RATE ↑

EARLY EXPIRATION

INTRATHORACIC PRESSURE ↑

THORACIC VEINS COMPRESSED → VENOUS PRESSURE↑

VENOUS RETURN MOMENTARILY↑ → VENOUS BARORECEPTORS +

HEART RATE ↑

REMAINDER OF EXPIRATION

CARDIAC OUTPUT ↑

ARTERIAL PRESSURE ↑

CAROTID BARORECEPTORS +

HEART RATE ↓

NOTES

DATE : _____   *OUTLINE #_____*

QUESTIONS

1._____
2._____
3._____
4._____
5._____

REVIEW QUESTIONS

I. Matching

A. Left ventricle
B. Semilunar valve
C. Atrioventricular valve
D. Right ventricle
E. Left atrium

__D__ 1. Resting systolic pressure in this chamber is approximately 25 – 35 torr
__B__ 2. Closure of this valve accounts for the second heart sound
__E__ 3. This valve closes because arterial pressure exceeds ventricular pressure
__C__ 4. This valve opens when atrial pressure exceeds ventricular pressure
__A__ 5. Resting systolic pressure in this chamber is approximately 125 – 150 torr

II. Multiple choice

__B__ 6. If pulmonary blood flow (from RV to LA) equals 6 units and pulmonary vascular resistance (from RV to LA) equals 6 units, then the difference between mean pressure at the beginning of the pulmonary trunk and the mean pressure at the ends of pulmonary veins would be: (remember F = P/R)

A. 1 unit
B. 36 units
C. 62 units
D. 5 units
E. 6 units

__E__ 7. According to the carotid baroreceptor reflex:

A. an increase in carotid arterial Pco_2 increases cardiac output
B. a decrease in carotid arterial Po_2 increases cardiac output
C. a decrease in carotid arterial pH increases cardiac output
D. a decrease in carotid pressure decreases cardiac output *opposite*
E. none

__C__ 8. The period of time in the cardiac cycle that immediately precedes the rapid ejection phase of ventricular systole is a very short phase called:

A. ventricular diastole
B. atrial systole
C. isovolumetric contraction
D. isovolumetric relaxation
E. diastasis

_D_ 9. During the cardiac cycle the papillary muscles and chorda tendinae function to:
  A. close semilunar valves
  B. open atrioventricular valves
  C. eject blood into arterial systems
  D. prevent eversion of the AV valves
  E. relay pacemaker signals to the ventricles

_E_ 10. Consider the relationship between flow (F), pressure (P), and resistance (R) to the flow of blood through a blood vessel and pick the correct statement from below:
  A. $P = F/R$ $F = P/R$
  B. $R = PF$
  C. if flow remains constant and resistance to flow increases, then pressure falls
  D. an increase in blood viscosity increases vascular resistance
  E. if pressure increases, flow will also increase unless resistance increases in proportion to pressure

_C_ 11. The bicuspid and tricuspid valves are closed:
  A. while the ventricles are in diastole _Relax_
  B. by the movement of blood from the atria to the ventricles
  C. when the ventricles are in systole _Contract_
  D. while the ventricles are filling
  E. while the atria are contracting and emptying

_B_ 12. Blood flow through the pulmonary circuit (vol./time) equals blood flow through the systemic circuit (vol./time) even though P for the pulmonary circuit is nearly five times less than P for the systemic circuit. The reason is:
  A. right and left ventricles function together
  B. pulmonary resistance is nearly five times less than systemic resistance
  C. pulmonary vessels cannot constrict
  D. oxygen must be absorbed by pulmonary blood at the same rate its being removed from systemic blood
  E. not evident in any of the preceding choices

_A_ 13. According to Poiseuille's law:
  A. blood flow through the artery is directly proportional to the pressure differential
  B. blood flow through an artery is directly proportional to the viscosity of the blood _Inverse_
  C. blood flow is directly proportional to the fourth power of the length of the blood vessel _Invers_
  D. blood flow is not influenced by the radius of the blood vessel. _yes_
  E. two of the preceding

_____14. The mitral and aortic valves simultaneously open during:
  A.   rapid ejection
  B.   isovolumetric contraction
  C.   isovolumetric relaxation
  D.   diastasis
  E.   no part of the cardiac cycle

_____15. The distance between successive R waves on Lead II EKG is 25 mm. Universal EKG recording speed is 25m/sec. The heart rate is:
  A.   60 BPM
  B.   70 BPM
  C.   80 BPM
  D.   100 BPM
  E.   75 BPM

_____16. All four intracardiac valves are closed during:
  A.   no part of the cardiac cycle
  B.   isovolumetric contraction
  C.   rapid filling of the ventricles
  D.   rapid ejection
  E.   AV node depolarization

_____17. An increase in blood pressure within the venae cavae and right atrium signifies an increase in venous return and reflexly:
  A.   excites the cardioinhibitory center
  B.   depresses cardioaccelerator center activity
  C.   increases heart rate
  D.   lowers systemic arterial pressure
  E.   - causes drowsiness

_____18. A lengthening of the RR interval suggests:
  A.   tachycardia
  B.   bradycardia
  C.   a slowing of pacemaker signal conduction to the ventricles
  D.   SA node damage
  E.   2:1 heart block

_____19. Which factor from Poiseuille's law inversely affects blood flow (volume per unit of time)?
  A.   hydrostatic pressure difference
  B.   blood vessel radius
  C.   blood viscosity
  D.   cross-sectional area of the blood vessel
  E.   molecular weights of plasma proteins

_____ 20. One of the following statements is false. Which one?
A. Resting cardiac output for an adult is about 5L/min.
B. Ventricular muscle has a long absolute refractory period.
C. At no time during the cardiac cycle are both atria and ventricles   in systole.
D. At no time during the cardiac cycle are both atria and ventricles in diastole.
E. In a normal heart the SA node and the AV node depolarize with the same frequency.

_____ 21. Compared to the systemic circulation, the pulmonary circulation:
A. has a higher pressure gradient from beginning to end
B. offers less resistance to the flow of blood
C. has no capillaries *Pulmonary capillaries pg559*
D. has a lower rate of blood flow through the circuit
E. two of the preceding

_____ 22. Which of the following signifies the time required for a pacemaker signal to spread from the SA node to the ventricular myocadium?
A. QT interval
B. ST segment
C. PR interval
D. QRS duration
E. RR interval

_____ 23. During the isovolumetric contraction phase of the cardiac cycle:
A. stoke volume increases
B. the AV node fires
C. the AV valves are closed
D. the semilunar valves are open
E. the ventricles are in diastole

_____ 24. Considering the events during one cardiac cycle, which of the following precedes depolarization of the atrioventricular node?
A. opening of the semilunar valves
B. closure of the atrioventricular valves
C. isovolumetric contraction
D. rapid filling of the ventricles
E. rapid ejection of the ventricles

_____ 25. The period of time that immediately follows the ejection phase of a cardiac cycle is called:
A. ventricular systole
B. isovlometric contraction
C. atrial systole
D. isovolumetric relaxation
E. diastasis

OUTLINE 15

LEARNING OBJECTIVES (√)

After reading the assigned pages in the textbook, and/or reading and performing related laboratory exercises, and listening to lecture and/or laboratory presentations, you should be able to:

☐1. Define systolic blood pressure and diastolic blood pressure and give the normal range of values for each pressure.

☐2. Use stroke volume (SV) and heart rate (HR) to define cardiac output (CO) and list the range of normal values for each.

☐3. Use an equation to explain the relation between cardiac output (CO), mean arterial pressure (MAP), and peripheral resistance (PR).

☐4. Explain the relation between stroke volume (SV), end-diastolic volume (EDV), and end-systolic volume (ESV).

☐5. Give two factors that influence EDV and explain how EDV changes in relation to each factor.

☐6. Give two factors that influence ESV and explain how ESV changes in relation to each factor.

☐7. Explain how increases and how decreases in heart rate and in stroke volume influence systemic blood pressure.

☐8. Define the cardiac ejection fraction and give the range of normal values.

☐9. Use an equation to define pulse pressure and explain the relation to pulse pressure of SV, HR, and PR.

☐10. Define mean arterial pressure (MAP), explain its significance, and give two formulae used to approximate MAP.

☐11. Outline two physiological mechanisms that serve to maintain MAP within its normal range.

☐12. Describe two ways local capillary blood flow can be altered to meet increased oxygen needs of local tissue.

☐13. Compare and contrast the pulmonary circuit with the systemic circuit as regards blood flow (liters per min.), pressure differentials from beginning to end, vascular resistance, and the regulation of flow through the circuit.

☐14. List normal values for four Starling forces operating at the systemic capillary, and explain how each force either favors or opposes filtration or absorption.

☐15. Explain the role of the lymphatic system in maintaining normal capillary dynamics.

☐16. List three things that alter capillary dynamics and lead to peripheral edema.

☐17. Explain why, in terms of Starling forces, there is no net filtration pressure in pulmonary capillaries, and the benefit of this.

NOTES

QUESTIONS

1._____
2._____
3._____
4._____
5._____

OUTLINE 15

LECTURER:______________ NOTES___________ DATE:______

I. SYSTEMIC BLOOD PRESSURE

A. Systolic vs. Diastolic

Diagram: graphic recording of systemic blood pressure

B. Influencing Factors

1. Cardiac output

4,800 mL   60   80

a. Cardiac output (CO) = Stroke volume (SV) x Heart rate (HR)

b. CO = Mean arterial pressure (MAP) /Peripheral resistance (PR)

80 mL

2. SV = End-diastolic volume (EDV) - End-systolic volume(ESV)

20

3. EDV factors

a. Atrial pressure

b. Ventricular compliance

4. ESV factors

a. Ventricular contractility

b. Ventricular afterload

5. Heart rate

Regulate out flow

Heart center of circulation

goal is to reach all cells w/ nutrients

hemodynamic - all blood flowing smooth to all cells

Pulse

Systolic pr. assures flow through vessels

(Pushes) Due to ventricle contraction

Veins + venules contain higher volume of blood

Diastolic - continuos flow after contraction or push

CO = cardic output

$CO = SV \times HR$

EX.

$P_1 = $ aortic pressure - $P_2 = $ vena cave

$S^{BP} - d^{BP} = PP$ "Normal" 120/80

$120 - 80 = 40 \, mmHG$

1._____
2._____
3._____
4._____
5._____

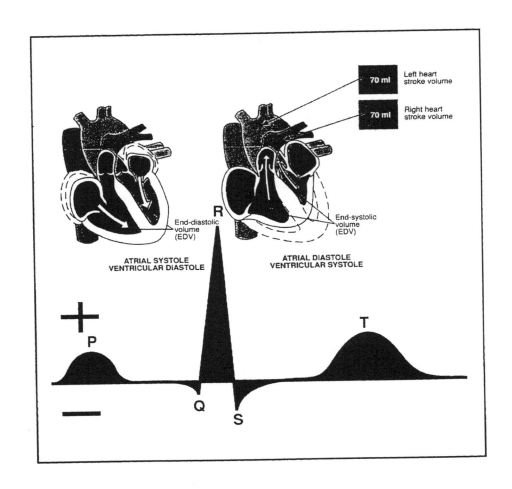

NOTES

DATE : _____   OUTLINE #_____

QUESTIONS

1. _____
2. _____
3. _____
4. _____
5. _____

6.   Ejection fraction = 55% - 65%
          Formula: (SV / EDV) x (100)

Normal Values	EDV	ESV	SV
Resting	110 -120 ml	40 - 50 ml	60 - 80 ml
Maximum	150 - 180 ml		120 - 160 ml
Minimum		10 - 20 ml	

C.   Pulse Pressure
    1.   Definition: systolic pressure – diastolic pressure = (pulse pressure)

    2.   Direct relation to SV: ↑SV ↑PP

    3.   Inverse relation to HR & PR: ↑HR, ↑PR ↓PP

D.   Mean Arterial Pressure
    1.   Measurement and computation
        a.   MAP = PP/3 + diastolic pressure
        b. $MAP = \dfrac{\text{Systolic Pressure} + 2 \times \text{Diastolic Pressure}}{3}$
    2.   Homeostatic mechanisms for maintenance
        a.   capillary fluid shift
        b.   renal adjustment of plasma volume
        c.   control of heart rate

II.   **BLOOD FLOW**
    A.   Autoregulation of Capillary Flow
        Diagram: tissue release of vasodilator substances

    B.   Autonomic Regulation of Arterioles
        Diagram: vasoconstrictor fibers to arterioles

$$map = \frac{sp + 2(DP)}{3}$$

$$100 = \frac{120 + 180}{3}$$

QUESTIONS

1._____
2._____
3._____
4._____
5._____

III. PULMONARY VS. SYSTEMIC PRESSURE AND FLOW
A. Pulmonary Flow = Systemic Flow

B. Pulmonary Pressure Differential < Systemic Pressure Differential

C. Pulmonary Resistance < Systemic Resistance

D. Regulation of Pulmonary Flow

IV. CAPILLARY DYNAMICS
A. Starling Forces: Filtration vs. Absorption
Diagram: systemic capillary and Starling forces

   1. Hydrostatic blood pressure

   2. Plasma protein osmotic pressure

   3. Tissue hydrostatic pressure

   4. Tissue protein osmotic pressure

B. Role of the Lymphatic System

C. Peripheral Edema: Heart Failure, Kwashiorkor, Liver Disease, etc.

D. Pulmonary vs. Systemic Capillary Dynamics
Diagram: pulmonary capillary and Starling forces

NOTES

Pg 614 Barorecyptors - stretch fires impulse to medulla
signals lower BP

Reduce Resistance by dialation of b vessels

$$MAP = CO \times TPR$$ BP depends on BV + R

Kidney cannot stand long term lack of blood

(the more Volume the more force) Frank starling
Hemmorage = fast Heart Rate /smaller quantity of blood
vascular constriction of Heart + skin does
not increase (local control)   not effected b symptetic

Legs, arms can survive w/o blood flow for hours

w/BP drop water pulled from tissue

| INCREASE SV First / then Tranfuse if necessary |

Hypertension = osmosis from tissue
61% blood in Veins /constrict veins ↑BP

"Reperfusion injury"
Heart Failure
Review Question ✓ pge 621

QUESTIONS

1._____
2._____
3._____
4._____
5._____

COMPARISIONS OF PULMONARY AND SYSTEMIC CIRCUITS

Pulmonary

1. Shorter circuit

2. Lower pressure circuit

3. Low resistance circuit

4. Blood flow 5/m   =

5. Velocity of Blood flow   =

6. Blood volume
   12% of total

9% in
Heart

7. Function
   oxygenation of blood
   (pick up O_2, release CO_2)

8. As CO $\uparrow$, BP $\uparrow$; Resistance decreases
   because vessels more distensible.

9. There is an absorption pressure
   all along the length of the
   pulmonary capillary.

10. No Carotid sinus, Aortic or Bain-
    bridge types of reflexes. Blood
    flow to parts of lung is determined
    largely by (O_2) and (CO_2) in various
    parts of the lung.
    $\downarrow CO_2 \uparrow O_2 \uparrow$ blood supply
    $\uparrow CO_2 \downarrow O_2 \downarrow$ blood supply

Systemic

1. Longer circuit

2. High pressure circuit

3. High resistance circuit

4. Blood flow 5L/min

5. Velocity of Blood flow

6. Blood volume
   79% of Total

61% in sys veins
11% in sys arteries
7% in capillaries and arterioles

7. Function
   Supply O_2 and nutrients to cells.
   Pick up CO_2 and wastes from cells.

8. As CO $\uparrow$, BP $\uparrow$; vessels offer more resistance
   because vessels are less distensible.

9. There is a net filtration pressure at the arterial
   end of cap and a net absorption pressure at
   the venous end of the capillary bed.

10. Regulating arterial BP we have carotid sinus
    reflex and aortic reflex. Regulating venous
    BP is Bainbridge reflex.

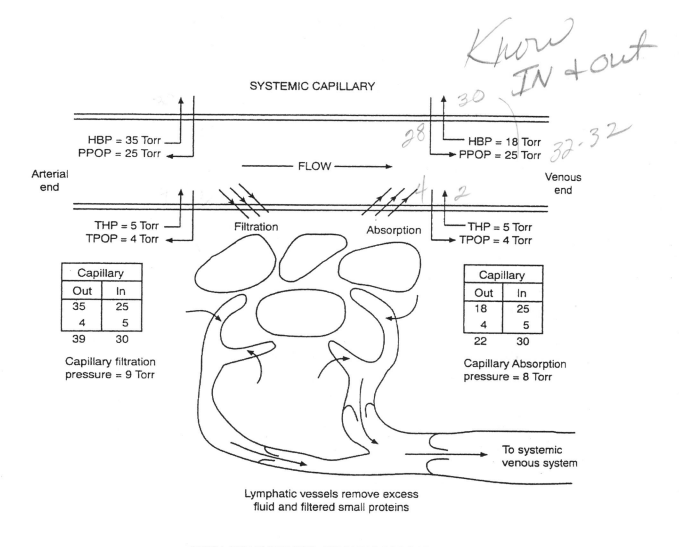

Know
IN + out

SYSTEMIC CAPILLARY

HBP = 35 Torr
PPOP = 25 Torr

Arterial end

FLOW

30
28

HBP = 18 Torr
PPOP = 25 Torr
32-32

Venous end

THP = 5 Torr
TPOP = 4 Torr

Filtration

4 2

Absorption

THP = 5 Torr
TPOP = 4 Torr

Capillary	
Out	In
35	25
4	5
39	30

Capillary filtration
pressure = 9 Torr

Capillary	
Out	In
18	25
4	5
22	30

Capillary Absorption
pressure = 8 Torr

To systemic
venous system

Lymphatic vessels remove excess
fluid and filtered small proteins

CAPILLARY DYNAMICS: STARLING FORCES

Peripheral edema: Swelling of tissues caused by excess fluid in the interstitial spaces.

Possible causes: 1. ↑ Venous HBP due to right heart failure

2. ↓ Arterial PPOP due to nutritional deficiency of protein (kwashiorkor)

3. ↑ TPOP due to blockage of lymphatic system due to parasitic worms, tumors, etc.

4. ↑ Capillary wall fragility due to nutritional deficiencies (eg. vitamin C and scurvy)

REVIEW QUESTIONS

I.   Matching

    A.   Resting end – diastolic volume

    B.   Resting end – systolic volume

    C.   Resting ejection fraction   *EDV – ESV*

    D.   Resting stroke volume

    E.   Resting cardiac output, HR = 70 BPM

Pg. 216 outline

    _C_ 1.   64%

    _A_ 2.   110 ml

    _D_ 3.   70 ml

    _B_ 4.   50 ml

    _E_ 5.   4.9 L/min

II.   Multiple choice

    _D_ 6.   As regards the systemic capillary, which of the following favors increased filtration and possibly the development of peripheral edema?

        A.   increased plasma colloidal osmotic pressure (PPOP)

        B.   decreased hydrostatic blood pressure (HBP)

        C.   increased interstitial hydrostatic pressure (THP)

        D.   increased interstitial colloidal osmotic pressure (TPOP)

        E.   none of the above

pg 6nd

    _E_ 7.   Given the following data, compute capillary absorption pressure: H.B.P. = 30 torr; P.P.O.P. = 28 torr; THP = 4 torr; TPOP = 2 torr.

        A.   8 torr

        B.   5 torr

        C.   3 torr

        D.   1 torr

        E.   Zero

Pg. 32/ outline

(handwritten calculation:)
$$\frac{30}{\underset{\overline{32}}{2}} \quad \Bigg| \quad \frac{28}{\underset{\overline{32}}{4}}$$

    _C_ 8.   Systolic blood pressure is 140 torr. Diastolic blood pressure is 70 torr. Mean arterial blood pressure is:

(handwritten:) $map = \frac{1}{3}(\underset{S}{140} - \underset{D}{70}) + \underset{+D}{70}$

        A.   78.3 torr   D.   108.3 torr

        B.   88.3 torr   E.   55 torr

        C.   93.3 torr

    _E_ 9.   Capillary filtration pressure will increase if the following pressure(s) decrease(s).

        A.   hydrostatic blood pressure (capillary)

        B.   plasma protein osmotic pressure (oncotic pressure)

        C.   tissue protein osmotic pressure

        D.   tissue hydrostatic pressure

        E.   two of the preceding

Pg. 32/ outline

_B_ 10. End-systolic volume (ESV):
  A. increases when ventricular contractility increases
  B. is always less than end-diastolic volume
  C. is the amount of blood in the ventricle just before the ventricle contracts
  D. is not affected by arterial pressure
  E. two of the preceding

_E_ 11. Since SV = EDV-ESV, left ventricular SV will be increased by all of the following except:
  A. increased atrial pressure
  B. increased ventricular pressure
  C. increased ventricular contractility
  D. increased ventricular compliance, i.e., an increase in ease with which the ventricle can be stretched
  E. increased aortic diastolic pressure

_E_ 12. Blood flow **through** the pulmonary circuit (vol./time) equals blood flow through **the systemic** circuit (vol./time) even though P for the pulmonary circuit is **nearly five times less** than P for the systemic circuit. The reason is:
  A. right and left ventricles function together
  B. pulmonary resistance is nearly five times greater than systemic resistance
  C. pulmonary vessels cannot constrict
  D. oxygen must be absorbed by pulmonary blood at the same rate its being removed from systemic blood
  E. pulmonary resistance is 5x less than systemic resistance

_C_ 13. If mean arterial pressure is 80 torr, diastolic pressure could be:
  A. 85 torr
  B. 90 torr
  C. 60 torr
  D. 87 torr
  E. 91 torr

$$MAP = \tfrac{1}{3}(Sys - D) + D$$
$$80 = \frac{Sys + 2(Dys)}{3}$$

_E_ 14. Given the following, compute cardiac output: H.R. = 70 BPM, EDV = 80 ml, ESV = 15 ml:
  A. 6650 ml   C. 5.6 L/min   E. 4550 ml/min
  B. 4.5 L   D. 4550 ml/beat

65×70

_D_ 15. If pulse pressure is 90 torr and mean arterial pressure is 90 torr, diastolic pressure is approximately:
  A. 180 torr
  B. zero
  C. 120 torr
  D. 60 torr
  E. 80 torr

$$90 = \frac{90}{3} + 60$$
MAP   PP   DP

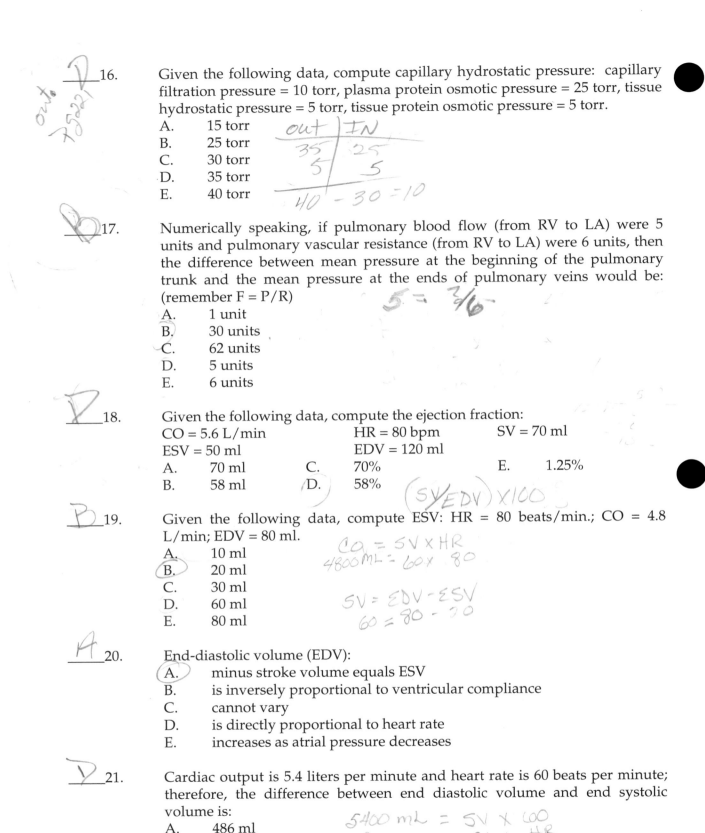

16. Given the following data, compute capillary hydrostatic pressure: capillary filtration pressure = 10 torr, plasma protein osmotic pressure = 25 torr, tissue hydrostatic pressure = 5 torr, tissue protein osmotic pressure = 5 torr.
    A.   15 torr
    B.   25 torr
    C.   30 torr
    D.   35 torr
    E.   40 torr

17. Numerically speaking, if pulmonary blood flow (from RV to LA) were 5 units and pulmonary vascular resistance (from RV to LA) were 6 units, then the difference between mean pressure at the beginning of the pulmonary trunk and the mean pressure at the ends of pulmonary veins would be: (remember F = P/R)
    A.   1 unit
    B.   30 units
    C.   62 units
    D.   5 units
    E.   6 units

18. Given the following data, compute the ejection fraction:
    CO = 5.6 L/min   HR = 80 bpm   SV = 70 ml
    ESV = 50 ml   EDV = 120 ml
    A.   70 ml   C.   70%   E.   1.25%
    B.   58 ml   D.   58%

19. Given the following data, compute ESV: HR = 80 beats/min.; CO = 4.8 L/min; EDV = 80 ml.
    A.   10 ml
    B.   20 ml
    C.   30 ml
    D.   60 ml
    E.   80 ml

20. End-diastolic volume (EDV):
    A.   minus stroke volume equals ESV
    B.   is inversely proportional to ventricular compliance
    C.   cannot vary
    D.   is directly proportional to heart rate
    E.   increases as atrial pressure decreases

21. Cardiac output is 5.4 liters per minute and heart rate is 60 beats per minute; therefore, the difference between end diastolic volume and end systolic volume is:
    A.   486 ml
    B.   4.8 liters
    C.   6 liters
    D.   90 ml
    E.   60 ml

224

_D__ 22.   Given the following data, compute heart rate:
SV = 60 ml/beat
EDV = 70 ml
ESV = 10 ml
CO = 5.4 L/min
HR = _____

(handwritten: $5400mL = 60 \times 90$ $CO = SV \times HR$)

A.   60 BPM
B.   70 BPM
C.   80 BPM
D.   90 BPM
E.   Cannot be computed without more data

_B__ 23.   Compared to the systemic circulation, the pulmonary circulation:
A.   has a higher pressure gradient from beginning to end
B.   offers less resistance to the flow of blood
C.   has a higher capillary filtration pressure
D.   has a lower rate of blood flow through the circuit
E.   two of the preceding

_C__ 24.   Which of the following would cause the largest increase in cardiac output?
A.   ↓EDV, ↑ESV, ↑HR
B.   ↑EDV, ↑ESV, ↓HR
C.   ↑EDV, ↓ESV, ↑HR
D.   ↓EDV, ↓ESV, ↓HR
E.   ↓EDV, ↓ESV, ↑HR

(handwritten: $CO = \overbrace{(EDV - ESV)}^{SV} \times HR$)

_D__ 25.   The principal reason for zero pulmonary filtration pressure in the normal lung is:
A.   plasma protein osmotic pressure is very low
B.   hydrostatic pressure in the pulmonary capillary is very high
C.   gases leave the blood and enter alveolar air
D.   pulmonary capillary pressure is very low
E.   pulmonary vascular resistance is very high

(handwritten:
$$\text{Filtration} = \frac{\overset{out}{HBP}}{TPOP} \Big/ \frac{\overset{In}{PPOP}}{THP}$$

$$\text{absorbtion} = \frac{\overset{out}{HBP}}{TPOP} \Big/ \frac{\overset{In}{PPOP}}{THP}$$
)

OUTLINE 16

LEARNING OBJECTIVES (√)

After reading the assigned pages in the textbook, and/or reading and performing related laboratory exercises, and listening to lecture and/or laboratory presentations, you should be able to:

☐1. List the principle parts of the upper and the lower divisions of the respiratory system.

☐2. List five basic functions of the upper division and two basic functions of the lower division.

☐3. Use an equation to define Boyle's law and explain its relation to the pressure, volume, and temperature of a gas.

☐4. Define Dalton's law, using the atmosphere as an example of a gas mixture, and explain what is meant by the term partial pressure.

☐5. Diagram the structural relation between the lungs, the thorax, visceral pleura, and parietal pleura, and label the base, root, and apex of the lung.

☐6. Use Boyle's law to explain changes in intrapulmonic pressure and intrapulmonic volume that occur during a normal resting respiratory cycle.

☐7. Draw a graph of intrapulmonic pressure vs. time for one complete resting respiratory cycle and use it to define negative and positive intrapulmonic pressures.

☐8. Explain why interpleural pressure remains negative (less than atmospheric pressure) during the normal respiratory cycle but intrapulmonic pressure does not.

☐9. Define pulmonary compliance and explain what is meant by increased pulmonary compliance and decreased pulmonary compliance.

☐10. Explain how and why pulmonary compliance changes when the lungs become less distensible in diseases such as black lung(coal miner's disease), or silicosis (grinder's disease).

☐11. Explain the role of surfactant in making the alveoli easier to expand during inspiration.

☐12. Explain why prematurely born infants sometimes have difficulty inflating their lungs.

☐13. List three factors that affect pulmonary compliance and explain how each factor affects compliance.

☐14. Define pneumothorax and atelectasis, and explain how and why changes in thoracic and lung volumes and pressures occur in unilateral pneumothorax.

☐15. Explain the significance of the difference between static and dynamic compliance.

NOTES

QUESTIONS

1._____
2._____
3._____
4._____
5._____

OUTLINE 16

I. **PRINCIPLE SUBDIVISIONS OF THE RESPIRATORY SYSTEM**
    A. Upper Division
        1. Nasal cavities
        2. Pharynx: nasopharynx, oropharynx, laryngopharynx
        3. Larynx

    B. Lower Division: The Respiratory Tree
        Diagram: respiratory airways, trachea to alveolus

        1. Trachea
        2. Primary bronchus   *Passage way*
        3. Lobar bronchus   anatomical dead space
        4. Segmental bronchus
        5. <u>Terminal bronchiole</u>
        6. Respiratory bronchiole
        7. Alveolar duct
        8. Alveolar sac   respiratory zone
        9. Alveolus

II. **BASIC FUNCTIONS OF THE UPPER DIVISION**
    A. Conduction *— Heat — transport*

    B. Cleansing *— sneeze - mucos*

    C. Humidification *— moisture & heat*

    D. Olfactory Support *— smell*

    E. Voice *larynx & air & sinus*

III. **BASIC FUNCTION OF THE LOWER DIVISION**
    A. <u>Gas Exchange</u> with the Blood: O_2 and CO_2

    B. Metabolic: acid – base regulation, ACE formation and blood
        pressure regulation, immunologic defense

228

NOTES

QUESTIONS

1._____
2._____
3._____
4._____
5._____

IV. MECHANICS OF PULMONARY VENTILATION

A. Gas Laws

ata constant
temp. pressure
or volume

1. Boyle's Law: PV = K   (T const.)

Example: (6) x (3) = 18

$$\downarrow \qquad \downarrow \qquad \text{pressure and volume vary inversely}$$

(9) x (2) = 18

Total Barometric
pressure = sum of
the partial pressure
of the individual gasses

2. Dalton's Law: Pt. = P_1 + P_2 + P_3 + etc.

Example: P_{atm} = P_{O_2} + P_{CO_2} + P_{N_2} + P_x

B. Structural Relationships: Lungs, Pleura, Thoracic Wall
Diagram: lungs, pleural cavities, thorax

1. Lung
2. Visceral pleura
3. Parietal pleura
4. Thoracic wall
5. Interpleural space (potential)
6. Pleural fluid

C. Muscles of Respiration
1. Inspiratory: external intercostals, diaphragm
2. Expiratory: internal intercostals, abdominal muscles

D. The Respiratory Cycle
Diagram: changes in intrapulmonic volume and pressure
Graph: Intrapulmonic Pressure vs. Time
1. Inspiration: intrapulmonic volume ↑ intrapulmonic pressure ↓

2. Expiration: intrapulmonic volume ↓ intrapulmonic pressure ↑

3. Changes in intrapleural (interpleural, intrathoraic) pressure
Diagram: intrapulmonic pressure (+) or (-)
interpleural pressure (-)

NOTES

OUTLINE #_____

QUESTIONS

1._____
2._____
3._____
4._____
5._____

V.   COMPLIANCE

A.   Definition:

$$\Delta V / \Delta P = \frac{\text{change in intrapulmonic volume}}{\text{change in interpleural pressure}}$$

B.   Factors Influencing Compliance
Diagram: lungs, thorax, alveolar sac

1.   Pulmonary and thoracic elasticity

2.   Surface tension - alveolar fluid - surfactant

3.   Airway resistance

4.   Pneumothorax and atelectasis
   a.   Definition
      1.   Pneumothorax

      2.   Atelectasis

   b.   Changes in lung & thorax
      Diagram: unilateral pneumothorax and atelectasis

C.   Static vs. Dynamic Compliance
Graph: measurement of static compliance and dynamic compliance

NOTES

DATE : _____

*OUTLINE #*_____

QUESTIONS

1. _____
2. _____
3. _____
4. _____
5. _____

REVIEW QUESTIONS

I.   Matching

     A. $\Delta V / \Delta P$ ~~notes~~

     B. $PV = K$ ~~outline~~

     C. $P_x = P_1 + P_2 + P_3$

     D. $F = \Delta P / R$

     E.   None of the preceding

_C_ 1.   Dalton's law *outline p230*

_A_ 2.   compliance *OL. 232*

_D_ 3.   a variation of Ohm's law *26p*

_B_ 4.   Boyle's law *OL p230*

_E_ 5.   the ideal gas law *636p*

II.   Multiple choice

_E_ 6.   During normal resting inspiration:

        (A.)   intrapulmonic pressure falls below atmospheric pressure

        B.   intrapulmonic volume decreases

        C.   intrapulmonic pressure increases

        (D.)   intrapulmonic volume increases

        E.   both answers (A) and (D) occur

PG 640

_A_ 7.   During inspiration (eupnea):

        A.   intrapulmonic pressure is never greater than atmospheric   pressure

        B.   intrapleural pressure is never less than atmospheric pressure

        C.   intrapleural pressure is never less than intrapulmonic pressure

        D.   answers A and B are correct but C is incorrect

        E.   answers A, B, and C are correct

_____ 8.   Which of the following increases during inspiration?

        A.   interpleural pressure

        B.   intrapulmonic pressure

        C.   intrathoracic pressure

        D.   intraalveolar pressure

        E.   none of the above

E

_B_ 9.   According to Boyle's law:

        A.   oxygen diffuses into pulmonary blood because of a higher partial pressure in the alveoli

        B.   air enters the lung during inspiration because intrapulmonic pressure is less than atmospheric as lung volume increases

        C.   systemic arterial blood carries more oxygen than pulmonary arterial blood

        D.   carbon dioxide is more soluble than oxygen in body water

        E. $PK = V$

Transpulmonary Pressure always
PG 639

_A_ 10.   Pulmonary compliance:
         A.   decreases as the lungs become maximally inflated
         B.   would not be affected by pneumothorax
         C.   does not change during maximal inspiration
         D.   increases as the lungs become maximally inflated
         E.   two of the preceding

_D_ 11.   Which of the following is not a function of the upper division of the respiratory system?
         A.   humidification of inspired air
         B.   warming of inspired air
         C.   cleansing of inspired air
         D.   gas exchange with the blood
         E.   olfaction

_____ 12.   At the beginning of inspiration, which force favors inflation of the lungs?
         A.   alveolar surface tension
         B.   elastic recoil of the lungs
         C.   elastic recoil of the thorax
         D.   airway resistance
         E.   intrapulmonic pressure

_E_ 13.   During normal resting expiration:
         A.   intrapulmonic pressure falls below atmospheric pressure
         B.   intrapulmonic volume increases
         C.   intrapulmonic pressure increases   P 640
         D.   intrapulmonic volume decreases
         E.   both answers (C) and (D) occur

_____ 14.   The function of surfactant is to:
         A.   increase surface area of the lung
         B.   reduce thickness of the pulmonary membrane
P 648   C.   reduce alveolar surface tension
         D.   increase alveolar surface tension
         E.   reduce pulmonary compliance

_C_ 15.   When a greater change in interpleural pressure is required to effect a former change in intrapulmonic volume:
         A.   the lungs have become more compliant
         B.   compliance of the thorax has increased
         C.   pulmonary compliance has decreased
         D.   airway resistance has decreased
         E.   pneumonthorax has occurred

_C_ 16.   In pneumothorax and complete atelectasis involving the right lung:
         A.   interpleural pressure would not be affected
         B.   intrapulmonic pressure would become negative
         C.   interpleural pressure would increase toward atmospheric
         D.   perfusion of the right lung would increase
         E.   perfusion of the right lung would not change

C 17. According to Dalton's law, if Gas A exerts a pressure of 500 torr and is a mixture of Gas B (15%), Gas C (70%), Gas D (10%) and Gas E, the partial pressure of Gas E is:

A. 50 torr
B. 75 torr
C. 25 torr
D. 350 torr
E. 250 torr

C 18. At the <u>end</u> of inspiration, which of the following pressures is lowest?

A. intra-alveolar
B. intrapulmonic
C. interpleural
D. atmospheric
E. none, they are all equal

C 19. During normal resting inspiration:

A. the diaphragm becomes more dome-shaped
B. the lungs become more compliant
C. alveolar surface tension is reduced
D. airway resistance helps inflate the lungs
E. the rib cage compresses the lungs

B 20. When a smaller change in intrapulmonic volume accompanies a given change in interpleural pressure:

A. airway resistance has decreased
B. pulmonary compliance has decreased
C. alveolar surface tension has decreased
D. thoracic muscles don't have to work as hard during inspiration
E. it is easier to breathe

NOTES

QUESTIONS

1._____

2._____

3._____

4._____

5._____

OUTLINE 17

LEARNING OBJECTIVES (√)

After reading the assigned pages in the textbook, and/or reading and performing related laboratory exercises, and listening to lecture and/or laboratory presentations, you should be able to:

☐1. Graph the relation between the following divisions of respiratory air: Tidal Volume (TV), Inspiratory Reserve Volume (IRV), Expiratory Reserve Volume (ERV), Residual Volume (RV), Expiratory Capacity (EC), Inspiratory Capacity (IC), Functional Residual Capacity (FRC), and Vital Capacity (VC).

☐2. Define pulmonary ventilation and alveolar ventilation, and use formulae to calculate normal values.

☐3. Explain why the rate of alveolar gas turnover is approximately 90 seconds at a resting respiratory rate of 12 - 15 cpm, and explain the significance of this fact as it relates to gas diffusion between the lung and the blood.

☐4. List four factors that influence the rate of gas diffusion between the lung and the blood, and explain the nature of the influence.

☐5. Outline the role alveolar oxygen and alveolar carbon dioxide play in the reflex control of pulmonary blood flow.

☐6. Define ventilation/perfusion ratio and calculate an ideal value for an adult, assuming that all parts of the lungs are equally ventilated and equally perfused.

☐7. Compare normal values for ventilation/perfusion ratios, from the apex to the base of the lungs in an erect adult, with the ideal value and explain how and why they differ.

☐8. Explain the effects of gravity on the pulmonary distribution of blood by comparing the ventilation/perfusion ratios in an erect adult with the ventilation/perfusion ratios in the same adult turned upside down.

☐9. Explain how knowledge about ventilation/perfusion ratios can be applied for the clinical benefit of an individual suffering from pulmonary disease.

NOTES

QUESTIONS

1. _____
2. _____
3. _____
4. _____
5. _____

OUTLINE 17

LECTURER:_____   NOTES_____   DATE:_____

I. DIVISIONS OF RESPIRATORY AIR
Graph: spirogram with volumes and capacities

A. Tidal Volume (TV)
B. Inspiratory Reserve Volume (IRV)
C. Expiratory Reserve Volume (ERV)
D. Residual Volume (RV)
E. Expiratory Capacity (EC) = TV + ERV
F. Inspiratory Capacity (IC) = TV + IRV
G. Functional Residual Capacity (FRC) = ERV + RV
H. Vital Capacity (VC) = IRV + TV + ERV

II. ALVEOLAR VENTILATION
A. Alveolar Ventilation vs. Pulmonary Ventilation
Computations:
Pulmonary Ventilation = (TV) x (RR)
Alveolar Ventilation = (TV – ADS) x (RR)

B. Rate of Turnover of Alveolar Gas
Diagram: alveolar expansion and contraction,
Graph: alveolar Po_2 vs. time

III. PULMONARY DIFFUSION
A. Structure of the Alveolar - Capillary Membrane

B. Influencing Factors
1. Partial pressure

2. Surface area of A - C membrane

3. Thickness of the A - C membrane

4. Solubility of gas in A - C membrane

NOTES

DATE : _____

*OUTLINE #*_____

QUESTIONS

1._____
2._____
3._____
4._____
5._____

IV. VENTILATION / PERFUSION RATIOS (V_A/Q)

A. Definition

1. $V_A = (TV - ADS) \times (RR)$

2. Q = RV cardiac output

B. Ideal Value and Normal Ratios
Diagram: erect lung with normal ratios from apex to base
Calculation:

C. Influencing Factors

1. Ventilation: transpulmonary pressures

2. Ventilation: airway resistance

3. Perfusion: alveolar O_2 and CO_2 levels

4. Perfusion: gravity

D. Clinical Application

NOTES

QUESTIONS

1._____
2._____
3._____
4._____
5._____

REVIEW QUESTIONS

I. Matching

A. Inspiratory Capacity
B. Expiratory Capacity
C. Vital Capacity
D. Functional Residual Capacity
E. Total Lung Capacity

E 1. IRV + TV + ERV + RV
B 2. TV + ERV
A 3. TV + IRV
D 4. ERV + RV
C 5. IRV + TV + ERV

II. Multiple choice

A 6. Ventilation/Perfusion ratios in the standing person are typically high, compared to the ideal, in the apex of the lung because:
A. ventilation of the apex is better than most other areas of the lung
B. gravity and vascular reflexes significantly reduce blood flow to the apex
C. the greatest changes in the intrapulmonic pressure occur in the apex
D. the clavicle (collarbone) interferes with elevation of the first rib
E. pulmonary vascular resistance is less than systemic vascular resistance

A 7. The maximal volume of gas that can be expired after a resting tidal expiration is called:
A. expiratory reserve volume
B. expiratory capacity
C. functional residual capacity
D. vital capacity
E. inspiratory reserve volume

E 8. Given the following data, determine alveolar ventilation: ADS: 100 ml; RR: 15 cpm; VC: 3500 ml; RV: 1200 ml; TV: 400 ml
A. 1300 ml
B. 300 ml/min
C. 5000 ml/min
D. 3000 ml/min
E. 4.5 L/min

$(TV - ADS) \times RR = AV$

$300 \times 15 = 4500 ml$

B 9. In pulmonary edema, pulmonary diffusion may decrease because:
- A. surface area of the pulmonary membrane increases
- B. thickness of the pulmonary membrane increases
- C. CO_2 is more soluble than O_2 in body water
- D. PO_2 decreases
- E. two of the preceding

C 10. If tidal volume is 600 ml, anatomical dead space is 150 ml, and alveolar ventilation is 9 liters/min, then respiratory rate is:
- A. 10 cycles/min
- B. 15 cycles/min
- C. 20 cycles/min
- D. 25 cycles/min
- E. 30 cycles/min

ADS

$(600 mL - 150) \times RR = 9000 mL$

$450 \times ? = 9000$

30

E 11. Which of the following is the largest?
- A. inspiratory reserve volume *3.1*
- B. tidal volume *.5L*
- C. expiratory reserve volume *1.2*
- D. residual volume *1.2*
- E. inspiratory capacity *3.6*

D 12. The volume of air remaining in the lungs at the end of a tidal (resting) expiration is:
- A. tidal volume *TV*
- B. expiratory capacity *EC = TV + ERV*
- C. residual volume *RV*
- D. functional residual capacity *FRC = ERV + RV*
- E. expiratory reserve volume *ERV*

B 13. Given the following data, compute alveolar ventilation:

TV = VC - (IRV + ERV)
TV = 4800 - (3300 + 1000)
TV = 4800 - 4300
TV = 500

ADS = 150 ml
VC = 4800 ml
RR = 15 CPM
IRV = 3300 ml
ERV = 1000 ml

$(TV - ADS) \times RR = AV$
$500 - 150 \times 15 = AV$
$350 \times 15 = AV \ mL/min$

1750
3500
5.25 mL/min

- A. 2250 ml/min
- B. 5.25 L/min (circled)
- C. 7.2 L/min
- D. 4500 ml/min
- E. 5250 ml

C 14. Functional residual capacity is the sum of:
- A. ERV + TV
- B. ERV + IRV
- C. RV + ERV
- D. TV + RV
- E. IRV + TV

B 15.   Which volume remains constant throughout a person's life?
  A.   vital capacity
  B.   expiratory reserve volume
  C.   inspiratory reserve volume
  D.   residual volume
  E.   none of the preceding

D 16.   The maximal volume of gas that can be expired after a maximal inspiration is called:
  A.   expiratory reserve volume
  B.   expiratory capacity
  C.   functional residual capacity
  D.   vital capacity
  E.   inspiratory reserve volume

E 17.   If alveolar ventilation is 4900 ml/min, anatomical dead space is 150 ml and respiratory rate is 14 cycles per minute, then tidal volume is:
  A.   475 cc
  B.   2100 cc
  C.   525 cc
  D.   707 cc
  E.   500 cc

$(TV - ADS) \times RR = AV$
$(TV - 150) \times 14 = 4900$
$500 -$

C 18.   The rate of pulmonary diffusion of gases:
  A.   decreases as pulmonary membrane thickness decreases
  B.   increases during inspiration and decreases during expiration
  C.   decreases as the surface area of the pulmonary membrane decreases
  D.   is directly proportional to the solubility of the gas in the pulmonary membrane
  E.   two of the preceding

D 19.   Which of the following is the sum of three primary lung volumes?
  A.   EC = TV + ERV
  B.   FRC = ERV + RV
  C.   IC = TV + IRV
  D.   VC = IRV + TV + ERV
  E.   RV = Residual volume

E 20.   Which of the following would you expect to increase during exercise as compared to rest?
  A.   inspiratory reserve volume
  B.   tidal volume
  C.   expiratory reserve volume
  D.   residual volume
  E.   answers (A), (B) and (C)

NOTES

DATE : _____   *OUTLINE #_____*

QUESTIONS

1._____

2._____

3._____

4._____

5._____

OUTLINE 18

LEARNING OBJECTIVES (√)

After reading the assigned pages in the textbook, and/or reading and performing related laboratory exercises, and listening to lecture and/or laboratory presentations, you should be able to:

☐1. Explain why oxygen diffuses from alveolar air into pulmonary capillary blood and give a normal value for the P_{O_2} differential.

☐2. List two modes of oxygen transport in the blood and explain their relative importance in the normal air-breathing adult.

☐3. Diagram the transport of oxygen from the lungs to body tissues and show the changes in oxygen partial pressures that drive oxygen diffusion.

☐4. Use a formula to define oxygen-carrying capacity of blood and give a normal adult value.

☐5. Define the oxygen extraction ratio and explain how its value changes from a state of body rest to a state of exercise.

☐6. Diagram the oxyhemoglobin dissociation curve and correctly label the ordinate, the abscissa, alveolar P_{O_2}, interstitial P_{O_2}, and the percent saturation of Hb at interstitial P_{O_2} and alveolar P_{O_2}.

☐7. Explain why a shift of the oxyhemoglobin dissociation curve to the right favors the unloading of oxygen from hemoglobin, and give two causative factors.

☐8. Explain why a shift of the oxyhemoglobin dissociation curve to the left favors the retention of oxygen by hemoglobin, and give two causative factors.

☐9. Use the oxyhemoglobin dissociation curve to explain hemoglobin's role as an oxygen buffer.

☐10. Diagram the transport of carbon dioxide from interstitial fluids to the lungs and show the changes in CO_2 partial pressure that drive CO_2 diffusion.

☐11. List three modes of carbon dioxide transport in the blood and the percentage of CO_2 carried in each mode.

☐12. Construct a diagram to explain the roles of the inspiratory center and the expiratory center in the genesis of the normal resting respiratory cycle.

☐13. Diagram the Hering-Bruer reflex and explain its normal function.

☐14. Explain what is meant by the direct effect and the indirect effect of elevated systemic blood P_{CO_2} on increasing alveolar ventilation.

☐15. Explain how and to what degree decreased pH and decreased P_{O_2} in systemic arterial blood stimulate alveolar ventilation.

NOTES

DATE : _____

OUTLINE #_____

QUESTIONS

1._____
2._____
3._____
4._____
5._____

OUTLINE 18

LECTURER:_____   NOTES_____   DATE:_____

I.   OXYGEN TRANSPORT
Diagram: alveolus, pulmonary capillary, heart, systemic capillary

    A.   Dissolved in the Blood (2%)

    B.   Oxyhemoglobin (98%)

    C. O_2 Carrying Capacity of Blood and Resting O_2 Extraction
Formula: Hb content (gm/dl) x (1.34 ml O_2/gm)

    D.   Oxyhemoglobin Dissociation Curve
Graph: % saturation Hb vs. P_{O_2}

        1.   Role of hemoglobin as an oxygen buffer

NOTES

QUESTIONS

1._____
2._____
3._____
4._____
5._____

II. CARBON DIOXIDE TRANSPORT
Diagram: cells; systemic capillary, heart, pulmonary capillary, alveolus

    A.   Dissolved in the Blood (10%)

    B.   Carbaminohemoglobin (20%)

    C.   Bicarbonate (70%)

III. REGULATION OF ALVEOLAR VENTILATION
    A.   Medullary Centers (primary)
    Diagram: medullary centers, lungs, thorax
        1.   Inspiratory center
        2.   Expiratory center

    B.   Pontine Centers (secondary)
        1.   Pneumotaxic center
        2.   Apneustic center

    C.   Hering - Bruer Reflex (inspiro – inhibitory)
    Diagram: stretch receptor reflex

    D.   Chemoreceptor Reflexes
    Diagram: carotid chemoreceptor reflexes
        1. P_{CO_2} (direct & indirect)

        2.   pH

        3. P_{O_2}

NOTES

DATE : _____   *OUTLINE #_____*

QUESTIONS

1._____

2._____

3._____

4._____

5._____

REVIEW QUESTIONS

I.   Matching

A.   Oxygen – carrying capacity of blood
B.   Saturation of hemoglobin in systemic venous blood (body at rest)
C.   Resting alveolar ventilation (ADS = 150 ml, TV = 400 ml, RR = 12)
D. ΔP_{CO_2}, pulmonary capillary to alveolus
E.   None of the above

_D_ 1.   5 torr
_B_ 2.   70%
_C_ 3.   3 L/min
_A_ 4.   20 ml/dl
_E_ 5.   4500 ml/min

II.   Multiple choice

_D_ 6.   Partial pressure of carbon dioxide is greatest in:
A.   interstitial fluid
B.   alveolar air
C.   pulmonary arterial blood
D.   intracellular fluid
E.   pulmonary venous blood

_E_ 7.   According to the oxyhemoglobin dissociation curve:
A.   the higher the P_{CO_2} the higher the Hgb saturation
B.   the lower the P_{CO_2} the lower the Hgb saturation
C.   at normal alveolar P_{O_2} most of the Hgb is saturated (more than 50%)
D.   at normal resting interstitial P_{O_2} most of the Hgb is saturated (more than 50%)
E.   two of the preceding

_B_ 8.   The Hering Bruer reflex:
A.   results in an increase in alveolar ventilation in response to an increased arterial P_{CO_2}
B.   limits inspiration
C.   depresses alveolar ventilation when arterial P_{O_2} is normal
D.   increases alveolar ventilation in response to a decreased arterial pH
E.   two of the preceding

_D_ 9.   The most powerful chemical influence on the regulation of alveolar ventilation is:
A.   the partial pressure of oxygen in systemic arterial blood
B.   the partial pressure of carbon dioxide in systemic venous blood
C.   the concentration of hydrogen ions in pulmonary arterial blood
D. P_{CO_2} in systemic arterial blood
E.   two of the preceding

___D___ 10. In pulmonary blood arriving at an alveolus:
  A.   bicarbonate is released into the alveolus
  B.   hydrogen ions are formed so that blood pH decreases
  C.   chloride ions move into the red cells (chloride shift)
  D.   blood pH tends to increase as carbon dioxide diffuses into the alveolus
  E.   two of the preceding

___E___ 11. Partial pressure of oxygen is greatest in:
  A.   pulmonary arterial blood
  B.   alveolar air
  C.   pulmonary venous blood
  D.   intracellular fluid
  E.   atmosphere

___B___ 12. According to the oxyhemoglobin dissociation curve:
  A.   an increase in blood pH (more alkaline) favors the unloading of oxygen from hemoglobin
  B.   when the body is at rest, less than one-third of available hemoglobin's oxygen-carrying capacity is used to deliver oxygen to tissues
  C.   when breathing air, hemoglobin saturation in pulmonary arterial blood is 98%
  D.   heme can carry carbon dioxide in addition to oxygen
  E.   the hemoglobin content of normal blood is 14 mg/dl

___B___ 13. Which of the following is true concerning regulation of alveolar ventilation:
  A.   the regulation of alveolar ventilation is a function of primary respiratory control centers located in the pons
  B.   the expiratory center normally does not play an active role (stimulating muscles) during resting respiration
  C.   an increase in P_{O_2} in systemic arterial blood is the most powerful stimulus to the respiratory centers
  D.   the respiratory centers tend to be stimulated by an increase in systemic arterial pH
  E.   AV = (TV)(RR)

___C___ 14. According to the oxyhemoglobin dissociation curve the change in the saturation of hemoglobin that occurs at rest when systemic arterial blood gives up oxygen in the systemic capillary is approximately:
  A.   0-10 percent
  B.   10-20 percent
  C.   20-30 percent
  D.   70 percent
  E.   100 percent

C 15. Which of the following account for most of the O_2 and CO_2 transport in the blood?
A. Oxyhemoglobin, dissolved CO_2
B. Reduced hemoglobin, carbaminohemoglobin
C. Oxyhemoglobin, bicarbonate
D. Oxyhemoglobin, carbaminohemoglobin
E. Dissolved O_2 and CO_2

B 16. If the oxygen carrying capacity of the blood is 20 vol. % and each gram of hemoglobin transports 1.34 ml of oxygen, the hemoglobin concentration is:
A. 13.4 gms/dl
B. 14.9 gms/dl
C. 26.8 gms/dl
D. 0.067 gms/dl
E. 19 gms/dl

D 17. CO_2 acts indirectly to stimulate respiratory center activity because it increases production of _____ in cerebrospinal fluid.
A. Ca^{++}
B. Na^+
C. K^+
D. H^+
E. O_2

D 18. The partial pressure of CO_2 in alveolar air:
A. is the same as the carbon dioxide content of the air being breathed
B. is the same as in pulmonary arterial blood
C. is less than the partial pressure of CO_2 in pulmonary venous blood
D. is relatively stable throughout a normal resting respiratory cycle
E. is greater than systemic arterial P_{CO_2} but less than atmospheric P_{CO_2}

NOTES

DATE : _____   *OUTLINE #_____*

QUESTIONS

1._____
2._____
3._____
4._____
5._____

OUTLINE 19

LEARNING OBJECTIVES (√)

After reading the assigned pages in the textbook, and/or reading and performing related laboratory exercises, and listening to lecture and/or laboratory presentations, you should be able to:

☐1. Outline the components of the urinary system and describe in one sentence the principal function of each component.
☐2. Define renal cortex, renal medulla, renal columns, and renal pyramids.
☐3. Outline the arterial supply and venous drainage of the kidney.
☐4. Diagram and label the structure of a cortical nephron, including the glomerulus, Bowman's capsule, proximal tubule, limbs and loop of Henle, distal tubule, collecting tubule, afferent and efferent arterioles, and the peritubular capillary plexus.
☐5. Define in one or two sentences what is meant by glomerular filtration, tubular reabsorption, and tubular secretion.
☐6. List normal values for four forces that favor or oppose glomerular filtration and explain the influence of each force on glomerular filtration.
☐7. Define effective filtration pressure (EFP) and glomerular filtration rate (GFR) and give normal resting adult values for each.
☐8. Describe the effects on GFR of afferent arteriole vasoconstriction and dilation, and efferent arteriole vasoconstriction and dilation.
☐9. Diagram the juxtaglomerular apparatus and explain how it operates in autoregulation of GFR.
☐10. Use a diagram to explain control of renal arterial blood pressure by the renin-angiotensin system.
☐11. List three mechanisms of tubular reabsorption and give an example of a substance reabsorbed by each mechanism.
☐12. List the solutes and the amount of water reabsorbed by the proximal tubule, the limbs of Henle, the distal tubule, and the collecting tubule.
☐13. Diagram the process of tubular secretion and give three examples of substances that can be secreted into the urine.
☐14. Define renal clearance of plasma, explain why it is a useful test of renal function, and give a formula for computing the plasma clearance rate of substance X .

NOTES

DATE : _____   *OUTLINE #_____*

QUESTIONS

1._____
2._____
3._____
4._____
5._____

OUTLINE 19

LECTURER:_____   NOTES_____   DATE:_____

I. **URINARY SYSTEM**
    Diagram: components of the urinary system
    A. Components
        1. Kidneys
        2. Ureters
        3. Urinary bladder
        4. Urethra

    B. General Functions

II. **RENAL STRUCTURE**
    Diagram: coronal section through kidney
    A. General Structure
        1. Renal cortex
        2. Renal medulla
        3. Renal pyramid
        4. Renal column

    B. Arterial Supply
        1. Renal artery
        2. Interlobar artery
        3. Arcuate artery
        4. Interlobular artery
        5. Afferent arteriole
        6. Glomerulus
        7. Efferent arteriole
        8. Peritubular plexus

    C. Venous Drainage

NOTES

DATE : _____   *OUTLINE #_____*

QUESTIONS

1._____
2._____
3._____
4._____
5._____

D. Nephron Structure
Diagram: the cortical nephron with blood supply
1. Glomerulus
2. Bowman's capsule
3. Proximal tubule
4. Descending Limb of Henle
5. Henle's Loop
6. Ascending Limb of Henle
7. Distal tubule
8. Collecting tubule

III. OVERVIEW OF NEPHRON FUNCTION
Diagram: linear nephron showing filtration, rebsorption, secretion
A. Glomerular Filtration

B. Tubular Reabsorption

C. Tubular Secretion

IV. GLOMERULAR FILTRATION
Diagram: glomerulus, Bowman's capsule, Starling forces
A. Filtration Forces
1. Glomerular pressure (HBP)
2. Plasma protein osmotic pressure (PPOP)
3. Capsular hydrostatic pressure (CHP)
4. Capsular osmotic pressure (COP)

B. Effective Filtration Pressure = HBP – PPOP – CHP + COP

C. Glomerular Filtration Rate (GFR)
1. Normal resting value: 125 ml/min

2. Regulation of GFR
Diagram: AA, glomerulus, EA, Bowman's capsule, JG apparatus
a. Afferent and efferent arteriolar diameter: vasoconstrictor fibers

NOTES

DATE : _____   *OUTLINE #_____*

QUESTIONS

1._____
2._____
3._____
4._____
5._____

          b.   Juxtaglomerular apparatus
                    Macula Densa (DCT)
                    JG Cells (AA)

D.   Renin-Angiotensin System
        (Renal Control of Renal Blood Pressure)
        Diagram: the renin angiotensin system

V.   TUBULAR REABSORPTION
   A.   Active Transport

   B.   Net Diffusion

   C.   Osmosis

VI. TUBULAR SECRETION
   A. H^+ and K^+

   B.   Other Substances

VII. RENAL CLEARANCE OF PLASMA (PLASMA CLEARANCE)
   A.   Definition
   B.   Computation:

$$\text{Clearance of X (ml/min)} = \frac{\text{Rate of Urinary Excretion of X (mg/min)}}{\text{Plasma Concentration of X (mg/ml)}}$$

Example: Clearance of urea = (12 mg/min) ÷ (0.2 mg/ml) = 60 ml/min

NOTES

DATE : _____   *OUTLINE #_____*

QUESTIONS

1._____
2._____
3._____
4._____
5._____

REVIEW QUESTIONS

I. Matching

A. Proximal tubule
B. Descending limb of Henle
C. Ascending limb of Henle
D. Distal tubule
E. Collecting tubule

_E_ 1. urea reabsorbed here contributes to high medullary osmotic pressure
_C_ 2. normally impermeable to water
_A_ 3. 100% of filtered glucose normally absorbed here
_D_ 4. reabsorption of Ca++ from the 15% remainder of filtered water
_D_ 5. reabsorption of water due to high osmotic gradient even when ADH levels in the kidney are significantly reduced

II. Multiple choice

_C_ 6. Capsular pressure = 3 torr, glomerular hydrostatic pressure = 55 torr, capsular osmotic pressure = 0, and plasma protein osmotic pressure = 27 torr. Glomerular filtration pressure is:
   A. 31 torr
   B. 10 torr
   C. 25 torr
   D. 81 torr
   E. none of the above

_C_ 7. ACE inhibitors reduce the:
   A. distal tubular reabsorption of glucose
   B. tubular transport maximum for glucose
   C. plasma concentration of angiotensin II
   D. gastric secretion of HCL
   E. release of bile from the gallbladder

_F_ 8. Renin release is stimulated by:
   A. increased blood pressure in afferent arterioles
   B. increased effective arterial blood volume
   C. increased sodium chloride transport by macula densa cells
   D. stimulation of renal sympathetic nerves
   E. an increase above normal plasma potassium level

_E_ 9. Obligatory water reabsorption:
   A. occurs in the proximal tubule
   B. occurs in the collecting tubule
   C. occurs deep in the renal medulla
   D. is not primarily controlled by ADH
   E. two of the preceding

A 10. Physiologically, GFR is reduced by:
- A. vasoconstriction of the afferent arterioles
- B. vasodilation of the afferent arterioles
- C. decreasing plasma protein concentration
- D. mild vasoconstriction of efferent arterioles
- E. two of the above

D 11. Approximately what percentage of filtered water is usually reabsorbed by the distal nephron?
- A. 70%
- B. 800%
- C. 100%
- D. 5%
- E. 25%

E 12. Assuming normal levels of blood glucose, where and how much of the filtered glucose is reabsorbed by the nephron?
- A. 50% proximal tubule, 50% distal tubule
- B. 100% distal tubule
- C. 5% limbs of Henle
- D. 70% proximal tubule
- E. 100% proximal tubule

D 13. At a plasma concentration of 0.2 mg/ml, the rate of urinary excretion of substance P is 10 mg/min. Plasma clearance of substance P by the kidneys is therefore:
- A. 2.0 mg/min
- B. 50 mg/min
- C. 50 mg/ml
- D. 50 ml/min
- E. none of the above

D 14. Which of the following has the greatest effect on increasing GFR?
- A. AA vasodilation, EA vasodilation
- B. EA vasodilation, AA vasoconstriction
- C. EA vasoconstriction, AA vasoconstriction
- D. AA vasodilation, EA vasoconstriction
- E. Vasoconstriction of interlobular artery

B 15. As ECF osmotic pressure rises above normal:
- A. the kidneys increase urine formation
- B. ADH release increases
- C. cells begin to osmotically gain water
- D. the distal nephron becomes impermeable to water
- E. salt retention increases

_B_ 16. The juxtaglomerular apparatus:
- A. increases GFR when tubular transport maxima are exceeded
- B. reduces glomerular blood pressure when solute concentration in distal tubular filtrate increases
- C. is part of the renal sympathetic nervous system
- D. consists in part of modified smooth muscle cells that form themacula densa
- E. secretes angiotensin to maintain renal blood pressure

_A_ 17. Which of the following would tend to cause ADH secretion to increase?
- A. cellular dehydration
- B. excessive drinking of water
- C. diabetes insipidus
- D. cellular hydration
- E. none of the above

_B_ 18. Capsular pressure = 10 torr, plasma protein osmotic pressure = 30 torr, glomerular hydrostatic pressure = 48 torr. The maximum that afferent arteriolar pressure could fall before glomerular filtration ceases is:
- A. 6 torr
- B. 7 torr
- C. 12 torr
- D. 9 torr
- E. 10 torr

_C_ 19. The juxtaglomerular apparatus is made up of cells from:
- A. the efferent arteriole and the proximal tubule
- B. the afferent arteriole and the proximal tubule
- C. the distal tubule and the afferent arteriole
- D. the collecting tubule and the peritubular plexus
- E. the afferent arteriole and the loop of Henle

_E_ 20. When ECF osmotic pressure increases:
- A. GFR increases
- B. GFR decreases
- C. ADH release decreases
- D. The urine becomes more dilute
- E. ADH stimulates the kidneys to retain more of the filtered water

NOTES

DATE : _____   OUTLINE #_____

QUESTIONS

1._____
2._____
3._____
4._____
5._____

OUTLINE 20

LEARNING OBJECTIVES (√)

After reading the assigned pages in the textbook, and/or reading and performing related laboratory exercises, and listening to lecture and/or laboratory presentations, you should be able to:

☐1. Define diuresis and diuretic, and explain why a diuretic may be useful in controlling some forms of systemic hypertension (high blood pressure).

☐2. List three types of diuretics, give an example of each type, and explain how each causes diuresis.

☐3. Diagram the renal control of ECF Na^+ and K^+ by way of the renin - angiotensin - aldosterone system.

☐4. Explain the role of the renin - angiotensin system in the maintenance of renal arterial blood pressure.

☐5. Define ACE inhibitor and explain why ACE inhibitors are often used to control systemic hypertension.

☐6. Define counter-current multiplication and counter-current exchange, and explain how they apply to the establishment and maintenance of the renal cortico-medullary osmotic gradient.

☐7. Explain what is meant by obligatory water reabsorption, where it occurs, and quantitatively how much of the filtered water is reabsorbed in an obligatory manner.

☐8. Define antidiuretic hormone (ADH), give its origin, and explain how it affects the distal nephron.

☐9. Use Greek derivatives to define the term diabetes, and explain the cause and effects of diabetes insipidus.

☐10. Draw a negative - feedback control system for the maintenance of extracellular fluid osmotic pressure. Include osmoreceptors in the hypothalamus, the posterior pituitary, ADH, and the kidney in the diagram.

NOTES

QUESTIONS

1._____
2._____
3._____
4._____
5._____

OUTLINE 20

I. DIURESIS AND DIURETICS

    A.   Definition

        1.   Diuresis

        2.   Diuretic

    B.   Types of Diuretics

        1.   Afferent arteriole dilators: xanthines, alcohol
             Diagram: AA vasodilation, ↑GFR

        2.   Osmotic diuretics: mannitol
             Diagram: renal tubules and ↓ water reabsorption

        3.   Metabolic inhibitors: thiazides

II. REGULATION OF ECF OSMOTIC PRESSURE

    A.   Solute Regulation (eg., Na^+, K^+)
        (Renin - Angiotensin - Aldosterone)
        Diagram: kidney, liver, brain, adrenal cortex pathways

NOTES

QUESTIONS

1._____
2._____
3._____
4._____
5._____

B. Solvent Regulation (ECF Water Content)
Diagram: a simple counter-current heat exchanger

1. Counter-current multiplication and exchange:
The corticomedullary osmotic gradient
Diagram: the nephron and counter-current mechanisms

2. Anti-diuretic hormone (ADH)

3. Diabetes insipidus: lack of ADH, water loss without sugar

4. Osmoreceptor control mechanisms
Diagram: negative feedback control system

NOTES

DATE : _____

OUTLINE #_____

QUESTIONS

1._____
2._____
3._____
4._____
5._____

REVIEW QUESTIONS

I. **Matching**
    A.   Collecting tubule
    B.   Proximal convoluted tubule
    C.   Ascending limb of Henle
    D.   Distal convoluted tubule
    E.   Descending limb of Henle

_D_ 1.   reabsorbs most of the remaining filtered water should a concentrated urine need to be excreted *D*

_C_ 2.   always impermeable to water, it adds solute to the interstitium helping to create and maintain an osmotic force essential for water reabsorption *C*

_B_ 3.   obligatory water reabsorption *B*

_E_ 4.   counter-current multiplication involves the ascending limb of Henle and this structure *E*

_A_ 5.   plays a major role in autoregulation of GFR *A*

II. **Multiple choice**

_C_ 6.   Which of the following directly stimulates the adrenal cortex to secrete a hormone that increases Na^+ reabsorption in the kidney?
    A.   renin
    B.   aldosterone
    C.   angiotensin II
    D.   water
    E.   A.C.E.

_D_ 7.   An increase in ECF osmotic pressure:
    A.   stimulates aldosterone secretion
    B.   stimulates renin release
    C.   stimulates angiotensin II formation
    D.   stimulates ADH secretion
    E.   causes diuresis

_D_ 8.   Diuresis can result from:
    A.   hypersecretion of ADH
    B.   efferent arteriolar vasodilatation
    C.   afferent arteriolar vasoconstriction
    D.   filtration of large solutes that are not reabsorbed
    E.   two of the preceding

D 9. Sodium chloride and _____ are important in establishing and maintaining the cortico-medullary osmotic gradient in the kidney.
A. glucose
B. amino acids
C. potassium
D. urea
E. ammonia

C 10. Counter-current multiplication involves the:
A. proximal and distal tubules
B. distal and collecting tubules
C. ascending and descending limbs of Henle
D. proximal tubule and vasa recta
E. collecting tubule and vasa recta

E 11. One of the following hormones elevates the plasma concentration of sodium by increasing renal tubular reabsorption. Which one?
A. ANP
B. ADH
C. STH
D. Glucagon
E. Aldosterone

D 12. Antidiuretic hormone (vasopressin):
A. release is accelerated when ECF osmotic pressure decreases below normal
B. inhibits renin release
C. is produced by the adenohypophysis
D. release is inhibited by ANF
E. decreases water permeability of the distal nephron

E 13. Approximately what percentage of filtered water is usually reabsorbed by the kidney tubules?
A. 1 percent
B. 20 percent
C. 15 percent
D. 70 percent
E. 99 percent

B 14. Which fluid has the highest osmotic pressure?
A. blood in afferent arteriole
B. blood in efferent arteriole
C. blood in renal artery
D. blood in renal vein
E. filtrate in proximal tubule

_E_ 15. Which of the following molecules plays a role in establishing the cortico-medullary osmotic gradient in the kidney?
- A. urea
- B. glucose
- C. amino acids
- D. sodium chloride
- E. two of the preceding

_B_ 16. Mannitol is an example of :
- A. a posterior pituitary hormone
- B. an osmotic diuretic
- C. a substance not easily filtered but easily reabsorbed
- D. an afferent arteriole vasoconstrictor
- E. an efferent arteriole vasoconstrictor

_C_ 17. Counter-current exchange involves the:
- A. collecting tubule and ascending limb of Henle
- B. descending limb of Henle and ascending vasa recta
- C. proximal tubule and distal tubule
- D. ascending and descending limbs of Henle
- E. afferent and efferent arterioles

_A_ 18. Renin
- A. is produced by cells that form the macula densa
- B. converts angiotensinogen to angiotensin II
- C. converts angiotensin I to angiotensin II
- D. inhibits the hypothalamic release of ADH
- E. stimulates the adrenal medulla

_E_ 19. The principal benefit of the corticomedullary osmotic gradient:
- A. is to be able to retain salt in the blood
- B. is to be able to reabsorb most of the filtered water and excrete a concentrated urine when necessary
- C. is to be able to regulate the release of hormones from the adrenal cortex
- D. is to be able to supply badly needed water to the renal tubular cells
- E. is to be able to excrete a very dilute urine when conditions dictate

_D_ 20. Diuretics that are metabolic inhibitors create diuresis by:
- A. blocking the action of ADH
- B. blocking water channels
- C. blocking electrolyte reabsorption
- D. vasoconstricting the afferent arteriole
- E. plugging up the collecting tubule

NOTES

$$PH = -log_{10}[H^+]$$

$[H^+]$ = hydrogen ion concentration

$$[H^+] = 10^{-8} \, gm/L \quad pH = 8$$

QUESTIONS

1. _____
2. _____
3. _____
4. _____
5. _____

OUTLINE 21
LEARNING OBJECTIVES (√)

After reading the assigned pages in the textbook, and/or reading and performing related laboratory exercises, and listening to lecture and/or laboratory presentations, you should be able to:

☐1. List three reasons why it is important to control extracellular hydrogen ion concentration.

☐2. Define the symbol pH and explain why pH numbers with values less than 7.00 indicate an acidic solution and pH numbers with values greater than 7.00 indicate an alkaline or basic solution.

☐3. Define the term acid and explain the difference between a strong acid and a weak acid. Give an example of each.

☐4. Define the term base and explain the difference between a strong base and a weak base. Give an example of each.

☐5. Use the bicarbonate buffer system to define a conjugate acid/base buffer pair and explain how each member of the pair buffers either a strong acid or a strong base.

☐6. List the major intracellular buffers and the extracellular buffers.

☐7. Distinguish physiological buffers from chemical buffers, and give two examples of physiological buffers.

☐8. Explain how the respiratory system can increase or decrease the pH of extracellular fluids by altering the ratio of CO_2 excretion rate/CO_2 production rate.

☐9. Draw diagrams to explain each of the three following renal mechanisms for controlling ECF pH: reabsorption of filtered HCO_3^-, excretion of fixed acids, and the excretion of ammonium.

☐10. List three common features of the renal mechanisms listed above.

☐11. List the normal range of systemic arterial blood values for each of the following: pH, Po_2, Hb sat., Pco_2, and $[HCO_3^-]$.

☐12. Define acidemia and alkalemia.

☐13. Define each of the following terms and list the primary blood indicator for each disturbance: respiratory alkalosis, nonrespiratory (metabolic) alkalosis, respiratory acidosis, nonrespiratory (metabolic) acidosis.

☐14. Explain the difference between compensation and correction for an acid-base disturbance.

☐15. Give an example of how a subject's systemic arterial pH, Pco_2, and $[HCO_3^-]$ can be collectively used to point to the subject's acid-base disturbance.

☐16. List causes, explain the compensations, and give examples of compensated blood gas data for each of the four primary disturbances of acid-base balance.

NOTES

DATE : ________ **OUTLINE #** 21

Less than 7.0 ph acid – 0 to 6.99 acidic

Neutral = 7

More than 7.0 ph alkaline base = 7.1 → 14 base(alkaline)

strong Acid

weak acid

strong base

weak base

Buffers

Intracellular Extracellular

physiological chemical

QUESTIONS

1.______________________________
2.______________________________
3.______________________________
4.______________________________
5.______________________________

OUTLINE 21

LECTURER:_____   NOTES_____   DATE:_____

I.   THE NEED TO CONTROL ECF [H⁺] *Extra cellular fluid Hydrogen*

II.   ACID AND BASES
    A.   Definition of pH = $-\log_{10}$ [H⁺]

    B.   The pH Scale

    C.   Strong vs. Weak Acids
        1.   Strong acid

        2.   Weak acid

    D.   Strong vs. Weak Bases
        1.   Strong base

        2.   Weak base

III. DEFENSES AGAINST ECF [H⁺] CHANGES
    A.   Chemical Buffer Systems
        1.   Buffer pairs and how they work
            a.   Buffering a strong acid

            b.   Buffering a strong base

DATE : _____   OUTLINE # 21

normal Range of systemic arterial blood values

pH   Po2   Hbsat.   Pco2   HCo3-

acidemia -

alkalemia -

Respiratory alkadosis -

Non-respiratory alkadosis - metabolic -

Respiratory acidosis -

Non-Respiratory acidosis - (metebolic)

acid-base disturbance - 4 primary
compensation   correction

QUESTIONS

1. _____
2. _____
3. _____
4. _____
5. _____

2. Major intra and extracellular buffers

B. Physiological Buffers
1. Respiratory control of ECF [H$^+$]

2. Renal control of ECF [H$^+$]
   a. Reabsorption of filtered HCO$_3^-$

   b. Excretion of fixed acids

   c. Excretion of ammonium

IV. **DEFINITIONS AND NORMAL BLOOD GAS VALUES**
A. Blood Gas Values
1. P$_{CO_2}$
2. P$_{O_2}$
3. Hb saturation
4. pH
5. [HCO$_3^-$]

B. Alkalemia

C. Acidemia

D. Alkalosis
1. Respiratory

2. Non-respiratory (metabolic)

NOTES

DATE : _____   OUTLINE # _21_

QUESTIONS

1._____
2._____
3._____
4._____
5._____

E. Acidosis
  1. Respiratory

  2. Non-respiratory (metabolic)

V. **DISTURBANCES OF ACID - BASE BALANCE**
  A. Compensation Processes
  1. Respiratory compensation (for a non-respiratory disturbance)

  2. Non-respiratory (metabolic) compensation (for a respiratory disturbance)

  B. Interpretation of Pco_2, $[H^+]$, and pH

  C. Respiratory Acidosis
  1. Causes

  2. Uncompensated blood gas data

  3. Compensations

  4. Compensated blood gas data

  D. Respiratory Alkalosis
  1. Causes

  2. Uncompensated blood gas data

  3. Compensations

  4. Compensated blood gas data

NOTES

OUTLINE # _21_

QUESTIONS

1._____
2._____
3._____
4._____
5._____

E.   Non-Respiratory (Metabolic)Acidosis
       1.   Causes

       2.   Uncompensated blood gas data

       3.   Compensations

       4.   Compensated blood gas data

F.   Non-Respiratory (Metabolic) Alkalosis
       1.   Causes

       2.   Uncompensated blood gas data

       3.   Compensations

       4.   Compensated blood gas data

DATE : _____

Norms

Variable	ART. Blood	Venous Blood
PH	7.35-7.45	7.35 - 7.45
Hgb Sat.	95% ↑	70 - 75%
PO_2	80-100torr	35-40 torr
$[HCO_3^-]$	22-26 ME/L	22-26 mB/L

Alkalemia — PH above 7.45
Alkalosis — process causing Alkemia

Acidemia — PH below 7.35
Acidosis — process causing Acidemia

Respiratory Acidosis — indicator PCO_2 at 45torr above in art. Blood
Respiratory Alkalosis — PCO_2 Below 35torr art. Blood

metabolic Acidosis — non-respiratory Process $[HCO_3^-]$ below
chemical 22-ME/L

metabolic Alkalosis — non-respiratory $[HCO_3^-]$ above
26 ME/L

QUESTIONS

1._____
2._____
3._____
4._____
5._____

289

PHYSIOLOGY

ELEMENTARY EVALUATION OF ACID-BASE BALANCE

I. **DEFINITIONS**

   A. The symbol pH

   pH is an abbreviation symbol used to indicate the concentration of hydrogen ions in a solution. By definition, pH stands for the negative logarithm (base 10) of the hydrogen ion concentration:

   $$pH = -\log_{10} [H^+]$$

   The brackets denote hydrogen ion concentration in grams/liter.

   Examples: If $[H^+] = 10^{-8}$ gm/liter, pH = 8

   If $[H^+] = 10^{-7}$ gm/liter, pH = 7

   pH = 7 is an easier way to say $[H^+] = .0000001$ gm/liter.

   pH = 8 is an easier way to say $[H^+] = .00000001$ gm/liter.
   Notice that a change of one whole pH number (eg., 7 to 8) represents a <u>tenfold </u>change in H^+ concentration.

   The pH scale extends from zero to fourteen (0-14). pH values less than 7.00 are indicative of an acid solution. pH values greater than 7.00 are indicative of an alkaline (basic) solution. The pH value of 7.00 is considered neutral (neither acidic nor basic).

pH :	<u>0 1 2 3 4 5 6</u>	<u>7</u>	<u>8 9 10 11 12 13 14</u>
reaction:	Acidic	Neutral	Basic
	(more H^+) *concentration*		(less H^+)

   B. Acid

   An **acid** is a substance that can donate (give up) free hydrogen ions to a solution. When placed into a solution, an acid increases the concentration of H^+ in the solution thereby lowering the pH of the solution. For example, when hydrochloric acid (HCl) is placed in solution it dissociates (ionizes) forming hydrogen ions (H^+) and chloride ions (Cl^-):

   HCl $\rightleftharpoons$ H^+ + Cl^-

   Strong acids are those which completely or almost completely dissociate into hydrogen ions and anions. Since they add large numbers of H^+ to solution, they greatly affect pH. HCl (hydrochloric acid) is an example of a strong acid.

   Weak acids are those which incompletely dissociate into hydrogen ions and anions. Since weak acids add small numbers of

H^+ to solution, they do not affect the pH of a solution as greatly as strong acids do. Carbonic acid (H_2CO_3), which dissociates into hydrogen ions (H^+) and bicarbonate ions (HCO_3^-), is an example of a weak acid:

$$H_2CO_3 \rightleftharpoons H^+ + HCO_3^-$$

C. Base

A **base** is a substance that can accept free hydrogen ions thereby removing them from solution. When placed into a solution, a base decreases the concentration of H^+ in the solution thereby increasing the pH of the solution. For example, when sodium hydroxide (NaOH) is added to an aqueous solution, the hydroxl ion binds H^+ forming water (H_2O):

$$NaOH + H^+ \rightleftharpoons Na^+ + H_2O$$

Strong bases readily accept free H^+ and therefore greatly influence pH by readily removing hydrogen ions from solution. Hydroxide is an example of a strong base.

Weak bases do not bind free H^+ as readily as strong bases do and therefore do not influence pH as markedly. Sodium bicarbonate ($NaHCO_3$) is an example of a weak base:

$$NaHCO_3 + H^+ \rightleftharpoons Na^+ + H_2CO_3$$

II. DEFENSE AGAINST ECF H^+ CHANGES

A. Chemical Buffers

Chemical buffers are substances that prevent marked changes in the pH of a solution when acids or bases are added to it. A buffer system is composed of a weak acid and its conjugate base, together called an *acid-base buffer pair*. An example is the extracellular bicarbonate buffer system, made up of <u>carbonic acid</u> and <u>sodium bicarbonate</u> ($H_2CO_3/NaHCO_3$).

As an illustration of how an acid-base buffer pair functions to minimize changes in pH when strong acids or strong bases are added to solution, consider the following two reactions:

1. $HCl + NaHCO_3 \longrightarrow H_2CO_3 + NaCl$
2. $NaOH + H_2CO_3 \longrightarrow NaHCO_3 + H_2O$

In reaction #1, hydrochloric acid (a strong acid), if added to solution without buffer would drastically decrease its pH. If added to a solution containing buffer (in this case $NaHCO_3$), the buffer reacts with the strong acid to produce a weak acid (in this case H_2CO_3), which does not greatly decrease pH, and a neutral salt (NaCl). The buffer therefore has minimized the decrease in pH resulting from the addition of strong acid.

In reaction #2, sodium hydroxide (a strong base), if added to solution without buffer, would drastically increase its pH. If added to a solution containing buffer (in this case H_2CO_3), the buffer reacts with the strong base to produce a weak base ($NaHCO_3$), which does not greatly increase pH, and neutral water. The buffer has therefore minimized the increase in pH resulting from the addition of a strong base.

Buffers *do not prevent changes in pH* when strong acids or strong bases are added to solution, they simply minimize the change. Buffers minimize the change by converting strong acids to weak acids, and by converting strong bases to weak bases.

Principal buffers in body fluids include:
1. *Bicarbonate Buffer System* - H_2CO_3/HCO_3^-
2. *Phosphate Buffer System* - $H_2PO_4^-/HPO_4^{2-}$
3. *Plasma Proteins* - Proteins are weak acids; combined with sodium they are weak bases.
4. *Hemoglobin* - When binding oxygen, hemoglobin gives up hydrogen ion to solution, acting as a weak acid. When releasing oxygen, hemoglobin binds hydrogen ion, acting as a weak base.

B.   Respiratory Control of ECF pH.
Respiratory rate and depth are influenced by changes in blood chemistry. Blood bathes chemoreceptors located in the aortic and carotid bodies. Chemoreceptors sense changes in arterial P_{CO_2}, H^+, and P_{O_2}, and send impulses to the respiratory centers in the medulla oblongata.

↑P_{CO_2}, ↑$[H^+]$, and/or ↓P_{O_2} stimulate rate and depth of respiration.

↓ P_{CO_2}, ↓ $[H^+]$, and/or ↑P_{O_2} tend to inhibit respiratory rate and depth.

Changes in arterial CO_2 content exert by far the strongest influence on respiratory drive. CO_2, by virtue of its solubility in body fluids, also is able to cross the blood-brain barrier and enter cerebrospinal fluid where it stimulates medullary respiratory center activity by directly affecting chemoreceptors on the ventral surface of the medulla oblongata. Indirectly, CO_2 in cerebrospinal fluid also stimulates medullary respiratory center activity by reacting with water to form carbonic acid which in turn dissociates to form H^+.

The H^+ stimulate the respiratory centers.

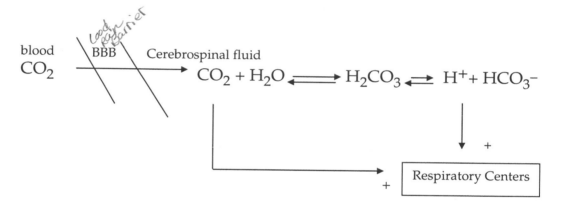

Increases above normal in systemic arterial P_{CO_2} have a maximum effect of increasing alveolar ventilation 10X normal. Increases above normal in systemic arterial $[H^+]$ have a maximum effect of increasing alveolar ventilation 5X normal. H^+ cannot cross the blood-brain barrier. They affect respiration by way of peripheral chemoreceptor reflexes (aortic and carotid). P_{O_2} changes are important in regulation of respiration mainly by changing the relative sensitivity of chemoreceptors to P_{CO_2}. If arterial P_{O_2} is very low, chemoreceptors become more sensitive to changes in arterial P_{CO_2}. Decreases below normal in systemic arterial P_{O_2} have a maximum effect of increasing alveolar ventilation 2X normal.

The respiratory system is able to participate in the regulation of body fluid pH by means of the following chemical reaction:

$$\overset{\uparrow \text{ alv. vent}}{\underset{\uparrow \text{metab. prod.}}{CO_2 + H_2O}} \overset{CA}{\underset{CA}{\rightleftarrows}} H_2CO_3 \rightleftarrows H^+ + HCO_3^-$$

Elimination of carbon dioxide from body fluids (via alveolar ventilation)at a rate faster than it is being produced (via metabolic production) would drive the reaction sequence to the left thereby reducing the amount of H^+ in the body fluids and raising the pH. This would occur in *alveolar hyperventilation*.

Elimination of carbon dioxide from the body fluids at a rate slower than it is being produced would drive the reaction sequence to the right thereby increasing the amount of H^+ in the body fluids and lowering the pH. This would occur in *alveolar hypoventilation*..

lower

The respiratory system supplements chemical buffers in controlling body fluid pH. If for example, body fluid pH is too high because of a nonrespiratory disturbance, the respiratory system may compensate by decreasing alveolar ventilation thereby retaining more CO_2 so as to reduce pH toward normal. If body fluid pH is too low because of a nonrespiratory disturbance, the respiratory system may compensate by increasing alveolar ventilation thereby removing CO_2 from the body faster than it is being produced. This tends to elevate pH back toward normal.

chemical reaction

Respiratory compensation for a nonrespiratory (metabolic) acid-base disturbance is usually incomplete. In metabolic alkalosis, the respiratory compensation is to decrease alveolar ventilation, allowing CO_2 to build up causing pH to fall toward normal, but as CO_2 level increases, respiratory center stimulation increases, thereby limiting the depression of alveolar ventilation as a compensatory response. In metabolic acidosis, the respiratory compensation is to increase alveolar ventilation, excessively removing CO_2, causing the pH to rise toward normal. However, as the CO_2 level decreases, respiratory center stimulation decreases, thereby limiting the increase of alveolar ventilation as a compensatory response.

C. Renal Control of Body Fluid pH
    The kidneys assist both the chemical buffer systems and the respiratory system in maintenance of body fluid pH by controlling both $[H^+]$ and $[HCO_3^-]$. Three basic renal mechanisms are involved: (1) the reabsorption of filtered bicarbonate, (2) production and

excretion of titratable acids, (3) excretion of ammonium salts. All three mechanisms are influenced by renal plasma P_{CO_2} (high P_{CO_2} speeds up, low P_{CO_2} slows down). All three mechanisms involve secretion of H^+ into the urine. All three mechanisms restore plasma HCO_3^-. Please refer to your textbook for diagrams of each of these mechanisms.

If body fluid pH is too low, the kidneys may compensate by increasing H^+ secretion and HCO_3^- reabsorption. If body fluid pH is too high, the kidneys may compensate by decreasing H^+ secretion and HCO_3^- reabsorption. The kidneys also assist in acid-base homeostasis by eliminating excess water and salts (end products of buffering).

III. DISORDERS OF ACID-BASE BALANCE

A. Normal Blood Gas Values (selected)

VARIABLE	ARTERIAL BLOOD	MIXED VENOUS BLOOD
pH	7.35 - 7.45	7.35 - 7.45
P_{O_2}	80 - 100 torr	35 - 40 torr
Hgb Sat.	95% or greater	70 - 75%
P_{CO_2}	35 - 45 torr	40 - 50 torr
$[HCO_3^-]$	22 – 26 MEq/L	22 - 26 MEq/L

Memorize

B. Definitions

Alkalemia: an alkaline condition of arterial blood, pH above 7.45
Alkalosis: the process causing alkalemia
Acidemia: an acid condition of arterial blood, pH below 7.35
Acidosis: the process causing acidemia

Respiratory Acidosis: a respiratory process causing acidemia. Indicator is P_{CO_2}, its relative value in arterial blood **above 45 torr**.

Respiratory Alkalosis: a respiratory process causing alkalemia. Indicator is P_{CO_2}, its relative value in arterial blood **below 35 torr**.

Metabolic Acidosis: a nonrespiratory process causing acidemia. Indicator is $[HCO_3^-]$, its value in arterial blood **below 22 MEq/L**.

Metabolic Alkalosis: a nonrespiratory process causing alkalemia. Indicator is $[HCO_3^-]$, its value in arterial blood **above 26 MEq/L**.

C. Evaluation of Acid-Base Disorders
1. Compensations for changes in body fluid pH by the chemical buffers and the kidneys are termed *metabolic compensation* . Compensations for changes in body fluid pH

by the chemical buffers and the respiratory system are termed *respiratory compensation*. Disturbances of primary respiratory origin cannot be compensated by respiratory means, but must be compensated for by metabolic means (and vice versa). *Complete compensation* means the ECF pH has been returned to normal range. *Incomplete compensation* means the ECF pH is returning toward normal but remains outside of the normal range.

2. In evaluating an acid-base disturbance, do not single out or consider separately any particular aspect of the disturbance in order to understand the consequences. For example, do not consider the carbonic anhydrase reaction separately but realize that several interrelated buffer reactions occur simultaneously.

3. Realize that any acid-base disturbance may be the sum of a primary disturbance and a compensatory or secondary disturbance, i.e., a subject may exhibit a primary metabolic alkalosis and subsequently develop secondarily, respiratory acidosis. Furthermore, notice that one needs to always establish the primary disturbances first in order to understand the compensating mechanisms.

4. Establish the primary disturbance. As pointed out previously, respiratory disturbances are assessed clinically by measuring the relative concentration of CO_2 (as Pco_2) in systemic arterial blood. Metabolic disturbances on the other hand are evaluated by measuring $[HCO_3^-]$ in systemic arterial blood.

ESTABLISHING THE PRIMARY DISORDER

5. In order to distinguish between a primary disturbance and a secondary compensation, compare values of $[HCO_3^-]$, Pco_2, and pH with normals. Remember $[HCO_3^-]$ is the indicator for nonrespiratory or metabolic disorders, and Pco_2 is the indicator for respiratory disorders. Look at each separately and decide whether the value is high, normal, or low, and then decide what the value means all by itself. When compensation is incomplete, the pH will be too high or too low and thus can be used to identify the primary disturbance because it (the pH) will point in the same direction as the indicator for the disturbance.

For example, subject A.B.C. has the following data:

$[HCO_3^-]$ = 34 MEq/L ↑ Met. alkalosis

Pco_2 = 48 torr ↑ Resp. Acidosis

pH = 7.46 ↑ Alkalemia

Examine the [HCO$_3^-$], then the P$_{CO_2}$, finally the pH.

High [HCO$_3^-$] = Metabolic (nonrespiratory) Alkalosis

High P$_{CO_2}$ = Respiratory Acidosis

High pH = Alkalemia

The pH points in the same direction as the bicarbonate concentration, therefore the primary disturbance is metabolic or nonrespiratory and the secondary or compensatory response is respiratory. Diagnosis: *Compensated Metabolic Alkalosis*.

6. If compensation is complete, the pH will be within the normal range. However, it will usually be at the low end of normal in compensated acidosis, and at the high end of normal in compensated alkalosis.

For example, subject D.E.F. has the following data:

$$[HCO_3^-] = 17 \text{ MEq/L} \quad \textit{Low met, Non-Res Acid}$$
$$P_{CO_2} = 30 \text{ torr} \quad \textit{Low Resp, Alka}$$
$$pH = 7.38 \quad \textit{Low normal}$$

Low [HCO$_3^-$] = Metabolic (nonrespiratory) Acidosis

Low P$_{CO_2}$ = Respiratory Alkalosis

pH = Normal, but at the low end of the normal range.

Diagnosis: *Compensated Metabolic Acidosis*

As you can see, in a completely compensated acid-base disorder, the pH is normal but both the [HCO$_3^-$] and the P$_{CO_2}$ are out of their normal range.

7. In a combined acid-base disturbance, the indicators and the pH all point in the same direction.

For example, subject G.H.I. has the following data:

$$[HCO_3^-] = 30 \text{ MEq/L} \quad \uparrow$$
$$P_{CO_2} = 30 \text{ torr} \quad \downarrow$$
$$pH = 7.6 \quad \uparrow$$

High [HCO$_3^-$] = Metabolic (nonrespiratory) Alkalosis

Low P$_{CO_2}$ = Respiratory Alkalosis

High pH = Alkalemia

Diagnosis: *Combined Metabolic and Respiratory Alkalosis*

The following table characterizes each of the four primary disturbances of acid-base balance in both compensated and uncompensated states. Consult your textbook for more detailed explanations of acid-base disorders.

EXAMPLES OF ACID-BASE ABNORMALITIES

Key: P.A. = Primary Abnormalities
Comp = Compensation

K = Kidney
L = Lung

Note: All values are arterial

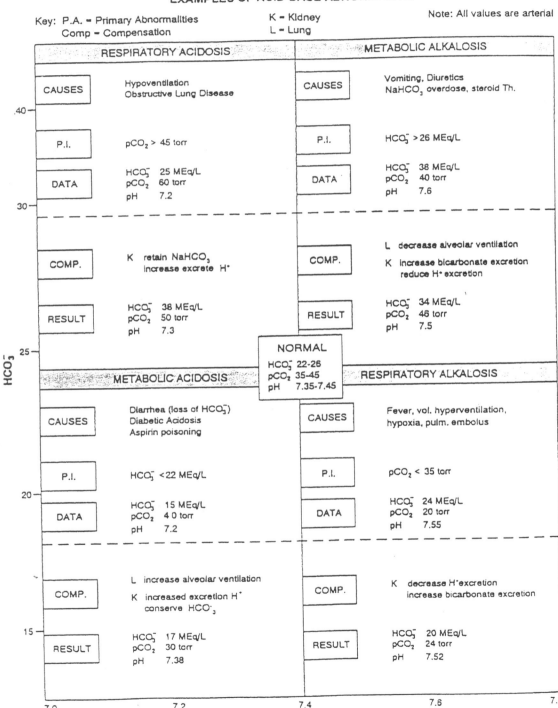

RESPIRATORY ACIDOSIS

CAUSES: Hypoventilation / Obstructive Lung Disease

P.I.: $pCO_2 > 45$ torr

DATA: HCO_3^- 25 MEq/L / pCO_2 60 torr / pH 7.2

COMP.: K retain $NaHCO_3$ increase excrete H^+

RESULT: HCO_3^- 38 MEq/L / pCO_2 50 torr / pH 7.3

METABOLIC ALKALOSIS

CAUSES: Vomiting, Diuretics / $NaHCO_3$ overdose, steroid Th.

P.I.: $HCO_3^- > 26$ MEq/L

DATA: HCO_3^- 38 MEq/L / pCO_2 40 torr / pH 7.6

COMP.: L decrease alveolar ventilation / K increase bicarbonate excretion reduce H^+ excretion

RESULT: HCO_3^- 34 MEq/L / pCO_2 46 torr / pH 7.5

NORMAL
HCO_3^- 22-26
pCO_2 35-45
pH 7.35-7.45

METABOLIC ACIDOSIS

CAUSES: Diarrhea (loss of HCO_3^-) / Diabetic Acidosis / Aspirin poisoning

P.I.: $HCO_3^- < 22$ MEq/L

DATA: HCO_3^- 15 MEq/L / pCO_2 4 0 torr / pH 7.2

COMP.: L increase alveolar ventilation / K increased excretion H^+ conserve HCO_3^-

RESULT: HCO_3^- 17 MEq/L / pCO_2 30 torr / pH 7.38

RESPIRATORY ALKALOSIS

CAUSES: Fever, vol. hyperventilation, hypoxia, pulm. embolus

P.I.: $pCO_2 < 35$ torr

DATA: HCO_3^- 24 MEq/L / pCO_2 20 torr / pH 7.55

COMP.: K decrease H^+ excretion increase bicarbonate excretion

RESULT: HCO_3^- 20 MEq/L / pCO_2 24 torr / pH 7.52

Y-axis: HCO_3^- (.40, 30, 25, 20, 15)

X-axis: pH (7.0, 7.2, 7.4, 7.6, 7.8)

NOTES

DATE : ________ *OUTLINE #* 21

Uncompensated	Compensated
M. Acid	M. ACIDOSIS
HCO_3 below norm	HCO_3 below normal
CO_2 norm	CO_2 below norm
PH below 7.35	PH normal
M. alkalosis	M. alkalosis
HCO_3 above 26 meq/L	HCO_3 — above norm.
CO_2 normal	CO_2 above norm
pH above 7.45	PH norm
Resp acid	Resp. Acid
CO_2 above normal 45 to	PCO_2 above N
HCO_3 norm	HCO_3 norm
PH below 7.35	PH Below norm
Resp alkalosis	Resp. Alka
CO_2 below norm 35 TORR	CO_2 above normal
HCO_3 norm	HCO_3 NOR
pH above 7.45	pH above norm

QUESTIONS

1. ______________________________
2. ______________________________
3. ______________________________
4. ______________________________
5. ______________________________

REVIEW QUESTIONS

I. Matching

Choice	Pco₂ mm Hg	HCO₃⁻ mEq/L	pH
A	30 ↓	14 ↓	7.29
B	50 ↑	40 ↑	7.53
C	60 ↑	31 ↑	7.34 ↓
D	60 ↑	23 N	7.21 ↓
E	25 ↓	24 N	7.55 ↑

C 1.   compensated respiratory acidosis
B 2.   compensated non-respiratory alkalosis
A 3.   compensated non-respiratory acidosis
D 4.   uncompensated respiratory acidosis
E 5.   uncompensated respiratory alkalosis

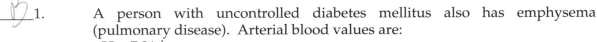

II. Multiple choice

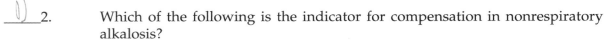

B 1.   A person with uncontrolled diabetes mellitus also has emphysema (pulmonary disease). Arterial blood values are:
pH = 7.21 ↓
Pco₂ = 60 torr ↑
Plasma (HCO₃⁻) = 20 mEq/L ↓
This subject has:
A.   nonrespiratory (metabolic) acidosis
B.   mixed respiratory and nonrespiratory acidosis
C.   mixed nonrespiratory (metabolic) and respiratory alkalosis
D.   respiratory alkalosis
E.   compensated nonrespiratory acidosis

D 2.   Which of the following is the indicator for compensation in nonrespiratory alkalosis?
A.   Pco₂ = 30 torr
B.   [HCO₃⁻] = 34 mEq/L
C.   [HCO₃⁻] = 20 mEq/L
D.   Pco₂ = 50 torr
E.   Pco₂ = 40 torr

B 3.   Which of the following is the indicator for compensation in respiratory acidosis?
A.   Pco₂ = 50 torr
B.   [HCO₃⁻] = 34 mEq/L
C.   [HCO₃⁻] = 24 mEq/L
D.   Pco₂ = 30 torr
E.   [HCO₃⁻] = 18 mEq/L

B 4. Which process in the kidney leads to generation of new bicarbonate to replenish depleted bicarbonate reserves?
- A. excretion of ammonia
- B. excretion of titratable acid
- C. excretion of sodium chloride
- D. reabsorption of urea
- E. A and B

 B 5. Which of the following could be associated with a value of 30 torr for systemic arterial plasma carbon dioxide partial pressure?
- A. gastric vomiting
- B. compensated nonrespiratory acidosis
- C. voluntary apnea
- D. uncompensated nonrespiratory alkalosis
- E. two of the above

 B 6. Aspirin poisoning, such as might occur when a person chronically ingests too much for relief from arthritis, may cause:
- A. uncompensated respiratory acidosis
- B. nonrespiratory acidosis
- C. nonrespiratory alkalosis
- D. respiratory alkalosis without renal compensation
- E. alkalemia

 B 7. Renal compensation for a respiratory acidosis includes:
- A. retention of ammonia
- B. increased formation of titratable acid in urine
- C. reduced reabsorption of filtered bicarbonate
- D. increased alveolar ventilation
- E. answers A and B

 B 8. All by itself, chronically elevated plasma bicarbonate (systemic arterial blood) suggests:
- A. respiratory acidosis
- B. nonrespiratory alkalosis
- C. respiratory alkalosis
- D. nonrespiratory acidosis
- E. compensation for a nonrespiratory acid-base disorder

E 9. In a hypothetical case of uncompensated nonrespiratory acidosis, which of the following could be true for systemic arterial blood?
- A. partial pressure of carbon dioxide = 25 torr
- B. bicarbonate concentration = 20 mEq/L
- C. PH = 7.39
- D. partial pressure of carbon dioxide = 60 torr
- E. bicarbonate concentration = 24 mEq/L

_____10.   The following data: $[HCO_3^-]$ = 18 mEq/L, Pco_2 = 38 torr, pH = 7.30, is indicative of:
   A.   compensated respiratory acidosis
   B.   uncompensated nonrespiratory acidosis
   C.   uncompensated respiratory alkalosis
   D.   compensated nonrespiratory alkalosis
   E.   compensated nonrespiratory acidosis

_____11.   In respiratory acidosis:
   A.   the kidneys compensate by conserving ammonia
   B.   plasma bicarbonate levels will slowly fall
   C.   systemic arterial Pco_2 will be low
   D.   renal tubular cells will secrete more H^+ into urine
   E.   two of the preceding

_____12.   With respect to normal values, during compensated respiratory acidosis, usually the:
   A. Pco_2 will be low, pH will be low
   B.   pH will be high, Pco_2 will be high
   C. $[HCO_3^-]$ will be high, Pco_2 will be low
   D. Pco_2 will be high, $[HCO_3^-]$ will be high
   E. Pco_2 will be low, $[HCO_3^-]$ will be low

NOTES

QUESTIONS

1._____
2._____
3._____
4._____
5._____

OUTLINE 22

LEARNING OBJECTIVES (√)

After reading the assigned pages in the textbook, and/or reading and performing related laboratory exercises, and listening to lecture and/or laboratory presentations, you should be able to:

☐1. Define the terms digestion and assimilation and give examples.
☐2. Diagram the alimentary canal as a long straight tube, label each major anatomical segment, and list the accessory organs associated with each segment.
☐3. List and explain the function of all sphincters of the alimentary canal.
☐4. Briefly explain the function of the myenteric plexus and Meissner's plexus (submucosal plexus).
☐5. Define the following propulsive and/or mixing movements of the alimentary canal: peristalsis, segmentation, haustration, mass movement.
☐6. Outline the four major effector components of the swallowing reflex.
☐7. Briefly describe three phases of salivary secretion.
☐8. Define and explain the significance of receptive relaxation.
☐9. List and briefly describe the functions of gastric secretions.
☐10. Explain how gastric motility and secretion are controlled by the autonomic nervous system.
☐11. List four hormones that control gastric motility and/or secretion, give their source, stimulus for release, and target cells.
☐12. Describe four factors that decrease the rate of gastric emptying.
☐13. Diagram the general structure of the duodenum, jejunum, and ileum, and draw a cross section of the jejunum to show the four tunics or coats of the intestinal wall, the myenteric plexus, and Meissner's plexus.
☐14. List digestive enzymes of intestinal mucosa origin, the substrates they act on, and the end-products of digestion.
☐15. Briefly outline neural and hormonal control of small intestinal motility and secretion.

NOTES

DATE : _____

OUTLINE # _22_

QUESTIONS

1._____
2._____
3._____
4._____
5._____

OUTLINE 22

LECTURER:_____   NOTES_____   DATE:_____

I. **GASTROINTESTINAL SYSTEM**
   A. General Functions
      1. Digestion

      2. Assimilation

   B. Structural Organization
      Diagram: alimentary canal and accessory organs
      1. Alimentary canal
         a. Oral cavity
         b. Pharynx
         c. Esophagus
         d. Stomach
         e. Duodenum
         f. Jejunum   } small intestine
         g. Ileum
         h. Cecum
         i. Ascending colon
         j. Transverse colon } large intestine
         k. Descending colon
         l. Sigmoid colon
         m. Rectum
         n. Anal canal

      2. Accessory organs
         a. Oral cavity: teeth, tongue, salivary glands
         b. Muscles of mastication
         c. Abdomen: liver, biliary system, exocrine pancreas

      3. Submucosal and myenteric plexi
         Diagram: cross section of gut

NOTES

DATE : _____   *OUTLINE #_____*

QUESTIONS

1._____
2._____
3._____
4._____
5._____

4. Sphincters
Diagram: location of sphincters
a. Upper esophageal
b. Lower esophageal (cardiac)
c. Pyloric
d. Ileocecal
e. Internal & external anal

II. ORAL CAVITY
A. Mastication

B. Salivary secretion
C. Swallowing reflex
Diagram: swallowing sequence
1. Elevation of soft palate
2. Relaxation of upper esophageal sphincter
3. Epiglottis moves to close off larynx
4. Pharyngeal constrictors contract

III. ESOPHAGEAL PERISTALSIS
Diagram: peristalsis in the esophagus

IV. STOMACH
A. General Structure
Diagram: external and internal gastric structure
1. Cardiac and pyloric sphincters
2. Greater and lesser curvatures
3. Corpora, fundus, and pylorus
4. Tunics
5. Gastric glands

B. Motor Functions
1. Receptive relaxation
2. Mixing movements and peristalsis

C. Gastric Secretions
1. Pepsinogen
2. HCl
3. Mucus

NOTES

DATE : _____   *OUTLINE #_____*

QUESTIONS

1._____
2._____
3._____
4._____
5._____

D. Regulation of Gastric Motility and Secretion
1. Neurogenic: ANS
2. Hormonal
   a. Gastrin
   b. Gastric inhibitory peptide
   c. Secretin
   d. Motilin
   e. Somatostatin
   f. Vasoactive intestinal polypeptide (VIP
   g. Enkephalin
   h. Bombesin

3. Rate of gastric emptying
   a. Degree of chyme fluidity
   b. Presence of irritants in duodenal chyme
   c. Presence of fat in duodenal chyme
   d. Distension of duodenum

V. **SMALL INTESTINE**
   Diagram: external and internal structure
   A. General Structure
   1. Duodenum
   2. Jejunum
   3. Ileum
   4. Ileo-cecal sphincter
   5. Glands

   B. Motor Functions
   Diagram: segmentation in small intestine

   1. Segmentation

   2. Peristalsis

NOTES

DATE : _____   *OUTLINE #_____*

QUESTIONS

1._____
2._____
3._____
4._____
5._____

C. Digestive Enzymes of SI Mucosal Origin
  1. Sucrase

  2. Maltase

  3. Lactase

D. Regulation of Motility and Secretion
  1. Neural control

  2. Hormonal control
    a. Gastrin

    b. Secretin

NOTES

DATE : _____   *OUTLINE #_____*

QUESTIONS

1._____
2._____
3._____
4._____
5._____

REVIEW QUESTIONS

I. Matching

A. Maltase
B. Lactase
C. Lipase
D. Carboxypeptidase
E. Sucrase

D 1. breaks peptide bonds
2. digests glycerides
A 3. digests disaccharide to form 2 molecules of glucose
B 4. digests disaccharide to form glucose and galactose
E 5. digests disaccharide to form glucose and maltose

II. Multiple choice

B 6. Gastrin:
  A. is a stomach enzyme
  B. stimulates pepsinogen secretion
  C. is released by sympathetic stimulation
  D. digests protein
  E. buffers HCL

E 7. The enterogastric reflex involves:
  A. inhibition of gastric secretion
  B. inhibition of colon motility
  C. stimulation of gastric secretion
  D. stimulation of pancreatic enzyme secretion
  E. relaxation of gastric sphincters

C 8. Which of the following does not control the passage of food material through the alimentary canal?
  A. upper esophageal sphincter
  B. cardiac sphincter
  C. sphincter of oddi
  D. pyloric sphincter
  E. ileo-caecal sphincter

C 9. The organ of greatest importance in providing the enzymes required for the digestion of food is the:
  A. parotid salivary gland
  B. stomach
  C. pancreas
  D. liver
  E. cecum

314

_D_ 10.   Distension of the distal small intestine relaxes this valve and distension of the proximal large intestine constricts it. The sphincter (valve) is the:
   A.   lower esophageal
   B.   upper esophageal
   C.   pyloric
   D.   ileocecal
   E.   external anal

Pg 698

_B_ 11.   Distension of the duodenum:
   A.   reflexly relaxes the cardiac sphincter
   B.   reflexly contracts the pyloric sphincter
   C.   reflexly stimulates gastric motility
   D.   reflexly inhibits jejunum motility
   E.   two of the preceding

_D_ 12.   Which of the following is (are) not present in saliva?
   A.   mucin
   B.   water
   C.   amylase
   D.   trypsin
   E.   two of the preceding

_C_ 13.   The rate of gastric emptying is:
   A.   primarily controlled by the amount of sugar in saliva
   B.   increased when fats enter the duodenum
   C.   inhibited by distension of the duodenum
   D.   inversely proportional to fluidity of chyme
   E.   two of the preceding

_D_ 14.   Receptive relaxation reflex involves:
   A.   duodenum
   B.   colon
   C.   ileum
   D.   stomach
   E.   esophagus

_D_ 15.   The enterogastric reflex involves:
   A.   inhibition of duodenal peristalsis by the stomach
   B.   stimulation of duodenal peristalsis by the stomach
   C.   inhibition of duodenal peristalsis by the colon
   D.   inhibition of gastric peristalsis by the duodenum
   E.   stimulation of gastric peristalsis by the duodenum

old book
pg 684

_A_ 16.   A hormone that stimulates secretion by the gastric parietal cell is:
   A.   gastrin
   B.   pepsin
   C.   gastric inhibitory peptide
   D.   secretin
   E.   calcitonin

_D_ 17.   Which of the following controls the passage of food material from the stomach?
- A.   upper esophageal sphincter
- B.   cardiac sphincter
- C.   sphincter of Oddi
- D.   pyloric sphincter
- E.   ileo-caecal sphincter

_B_ 18.   An enzyme that breaks down proteins into proteoses, peptones and polypeptides:
- A   gastrin
- B.   pepsin
- C.   maltase
- D.   lipase
- E.   sucrase

_B_ 19.   This hormone stimulates gall bladder contraction and also pancreatic secretion of digestive enzymes. It is called:
- A.   gastrin
- B.   cholecystokinin
- C.   gastric inhibitory peptide
- D.   secretin
- E.   glucagons

_C_ 20.   This motor function is unique to the small intestine. It is:
- A.   haustration _Large_
- B.   peristalsis _Both_
- C.   segmentation
- D.   mass movement _Both_
- E.   receptive relaxation _Stomach_

NOTES

DATE : _____   *OUTLINE #_____*

QUESTIONS

1._____
2._____
3._____
4._____
5._____

OUTLINE 23

LEARNING OBJECTIVES (√)

After reading the assigned pages in the textbook, and/or reading and performing related laboratory exercises, and listening to lecture and/or laboratory presentations, you should be able to:

☐1. Draw the general compound tubular-alveolar structure of the exocrine pancreas, label acinar cells and intercalated duct cells, and list the general type of secretion of each cell.

☐2. List the major secretions of the exocrine pancreas and describe their digestive functions.

☐3. Explain how exocrine pancreatic secretion is controlled by the autonomic nervous system.

☐4. List the source and target cells of secretin and cholecystokinin and explain the usual chemical stimuli for their release.

☐5. Diagram the general structure of the liver. Include the following components of the lobule: hepatic cells, sinusoids, Kupfer cells, central vein, portal triad.

☐6. Diagram the hepatic portal circulation and show how venous blood from the absorptive organs passes through the liver before entering general systemic venous circulation.

☐7. List five general yet distinct functions of the liver.

☐8. Diagram the biliary duct system and explain neural and hormonal control of bile secretion.

☐9. Outline the secretory and absorptive roles of the large intestine.

☐10. Define the following gastrointestinal reflexes, giving where possible, the usual stimulus that initiates the reflex, and the origin and target cells of the reflex: vomiting, gastrocolic, enterogastric, duodenocolic, peritoneal, mucosal, and defecation.

☐11. Explain the basic chemical organization of carbohydrates, using the monosaccharide as a basic component, and explain how ingested carbohydrates are digested and assimilated.

☐12. Diagram a generic amino acid, define a peptide bond, peptide, proteose, peptone, and protein, and explain how ingested proteins are digested and assimilated.

☐13. Define glycerol and fatty acid, and show how they are linked to form fats.

☐14. Briefly outline the processes of lipid digestion and assimilation, including the roles of both the portal blood and the intestinal lymphatics in lipid transport.

NOTES

DATE : _____   OUTLINE #_____

QUESTIONS

1._____
2._____
3._____
4._____
5._____

OUTLINE 23

LECTURER:_____   NOTES_____   DATE:_____

I. **EXOCRINE PANCREAS**
    Diagram: pancreas location and general structure
    A. General Structure
        1. Acinar cells

        2. Intercalated duct cell

    B. Secretions
        1. <u>Proteolytic</u>:trypsinogen, chymotrpsinogen, procarboxypeptidase

        2. <u>Amylolytic</u>: amylase

        3. <u>Lipolytic</u>: lipase, phospholipase, cholesterol esterase

        4. <u>Nucleolytic</u>: ribonuclease, deoxyribonuclease

        5. $NaHCO_3$: buffer

    C. Control of Secretions
        1. Neural
        2. Hormonal
            a. Secretin

            b. Cholecystokinin (CCK)

II. **LIVER AND THE BILIARY SYSTEM**
    Diagram: the liver and biliary duct system
    A. General Hepatic Structure
        1. Lobes
        2. Hepatic lobule
            a. Hepatic cells
            b. Sinusoids
            c. Kupfer cells
            d. Portal triad
            e. Central vein

NOTES

DATE : _____

OUTLINE #_____

QUESTIONS

1._____

2._____

3._____

4._____

5._____

B. Hepatic Portal Circulation
Diagram: blood flow to and through the liver

C. Major Hepatic Functions

D. Bile Formation and Secretion

E. Regulation of Bile Secretion
1. Neural

2. Hormonal
   a. Cholecystokinin

   b. Secretin

III. LARGE INTESTINE
Diagram: the external structure of the large intestine
A. General Structure
1. Cecum (+Vermiform Appendix)
2. Colon
   a. Ascending
   b. Transverse
   c. Descending
   d. Sigmoid
3. Rectum
4. Anal Canal

NOTES

DATE : _____   *OUTLINE #*_____

QUESTIONS

1._____
2._____
3._____
4._____
5._____

B. Secretions
    1. Mucus
    2. Potassium
    3. Bicarbonate
C. Absorption: Vitamins, Salt, Water

IV. GASTROINTESTINAL REFLEXES
A. Vomiting

B. Gastrocolic

C. Enterogastric

D. Duodenocolic

E. Peritoneal

F. Mucosal

G. Defecation

V. DIGESTION AND ASSIMILATION
A. Carbohydrates
    1. Structure

    2. Enzymatic Digestion

    3. Assimilation

NOTES

DATE : _____   *OUTLINE #_____*

QUESTIONS

1._____
2._____
3._____
4._____
5._____

B. Proteins
    1.   Structure

    2.   Enzymatic digestion

    3.   Assimilation

C. Lipids
    1.   Structure

    2.   Enzymatic digestion

    3.   Assimilation

D. Electrolytes and Water

NOTES

DATE : _____   *OUTLINE #_____*

QUESTIONS

1._____
2._____
3._____
4._____
5._____

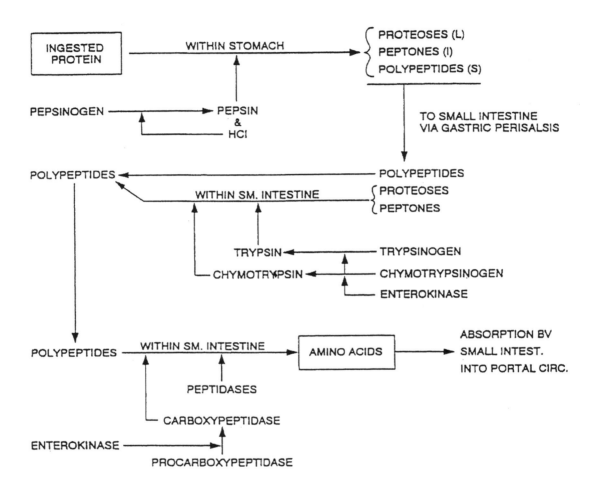

INGESTED PROTEIN → WITHIN STOMACH → ⎧ PROTEOSES (L)
⎨ PEPTONES (I)
⎩ POLYPEPTIDES (S)

PEPSINOGEN → PEPSIN & HCl

TO SMALL INTESTINE VIA GASTRIC PERISALSIS

POLYPEPTIDES ← POLYPEPTIDES

WITHIN SM. INTESTINE ⎧ PROTEOSES
⎨ PEPTONES

TRYPSIN ← TRYPSINOGEN
CHYMOTRYPSIN ← CHYMOTRYPSINOGEN
ENTEROKINASE

POLYPEPTIDES → WITHIN SM. INTESTINE → AMINO ACIDS → ABSORPTION BV SMALL INTEST. INTO PORTAL CIRC.

PEPTIDASES

CARBOXYPEPTIDASE

ENTEROKINASE →
PROCARBOXYPEPTIDASE

BASIC SET OF 20 AMINO ACIDS

NONESSENTIAL		ESSENTIAL	
ALANINE	GLUTAMINE	ARGININE	METHIONINE
ASPARAGINE	GLYCINE	HISTIDINE	PHENYLALANINE
ASPARTATE	PROLINE	ISOLEUCINE	THREONINE
CYSTEINE	SERINE	LEUCINE	TRYPTOPHAN
GLUTAMATE	TYROSINE	LYSINE	VALINE

NOTES

QUESTIONS

1._____
2._____
3._____
4._____
5._____

REVIEW QUESTIONS

I.   Matching

A.   Pepsin
B.   Carboxypeptidase
C.   Lipase
D.   Maltase
E.   Amylase

 1.   splits protein into peptide fragments *A*
 2.   digests disaccharides into monosaccharides *D*
____ 3.   digests glycerides to glycerol and fatty acids *C*
 4.   splits amino acids from peptides *B*
 5.   digests starch *E*

II.   Multiple choice

____ 6.   Which of the following elevates hepatic portal blood glucose the most?
A.   lactase
B.   sucrase
C.   maltase
D.   amylase
E.   insulin

____ 7.   The mucosal reflex involves:
A.   inhibition of gastric motility
B.   stimulation of duodenal secretion
C.   stimulation of salivary secretion
D.   inhibition of gastric secretion
E.   stimulation of colon motility

____ 8.   Which substance promotes the release of bile from the gallbladder?
A.   gastrin
B.   secretin
C.   norepinephrine
D.   cholecystokinin
E.   bicarbonate

____ 9.   Which enzyme degrades protein to peptones, proteoses, and polypeptides?
A.   trypsin
B.   peptidase
C.   chymotrypsin
D.   pepsin
E.   HCl

_B_ 10. The various process of the gastrointestinal system are directly associated with certain types of gastrointestinal cells. Which of the following association is NOT valid?

A. motility — esophageal smooth muscle cells
B. secretion — gallbladder epithelial cells
C. digestion — pancreatic acinar cells
D. absorption — small intestinal epithelial cells
E. absorption — gastric mucosal cells

_E_ 11. One of the following enzymes functions best when the pH of its environment is very low. Which one?

A. trypsin
B. amylase
C. carboxypeptidase
D. sucrase
E. pepsin   Less than 5

_D_ 12. Acids in duodenal chyme promote the release of this hormone:

A. sodium bicarbonate
B. ACTH
C. gastrin
D. secretin
E. ptyalin

_C_ 13. Peptidases:

A. split fatty acids from glycerides
B. are secreted by gastric, intestinal, and pancreatic cells
C. break bonds between amino acids pg 47-48
D. are found in saliva
E. two of the preceding pg 698 old Book

_B_ 14. The smallest end-product resulting from lipase activity is (are):

A. ketoacids
B. glycerol
C. monosaccharides
D. amino acids
E. triglycerides

_C_ 15. Secretin:

A. inhibits large intestine secretion
B. stimulates gastric peristalsis
C. stimulates release of $NaHCO_3$ by pancreas
D. stimulates release of pancreatic insulin
E. is the same hormone as pancreozymin

_C_ 16. Which of the following splits fatty acids from triacylglycerol?
    A.   amylase
    B.   sucrase
    C.   lipase
    D.   ribonuclease
    E.   lactase

_C_ 17. Which of the following are absorbed primarily into the intestinal lymphatics?
Pg 718-19
    A.   monosaccharides
    B.   disaccharides
    C.   glycerides → chylomicrons
    D.   dipeptides
    E.   amino acids

_D_ 18. Which of the following is an endopeptidase?
    A.   pepsin
    B.   maltase
    C.   amylase
    D.   enterokinase
    E.   HCL

_B_ 19. Motility of the colon is:
    A.   stimulated by norepinephrine
    B.   increased by gastric distension
    C.   decreased by duodenal distension
    D.   inhibited by the vagus
    E.   two of the preceding

_C_ 20. Pancreatic lipase:
    A.   digests peptides
    B.   requires secretin for release
    C.   splits triglyceride into monoglyceride and free fatty acids
    D.   forms chylomicrons
    E.   two of the above

_D_ 21. Gluconeogenesis:
    A.   occurs in skeletal muscle when glycogen is depleted
    B.   is stimulated by insulin in order to maintain blood sugar levels
    C.   occurs in the liver primarily after ingesting a meal high in carbohydrate content
    D.   involves the formation of glucose from non-carbohydrate sources by liver and kidney
    E.   two of the preceding

_D_ 22. From which part of the alimentary canal is the greatest amount of water absorbed?
A. stomach
B. esophagus
C. large intestine
D. small intestine
E. pharynx

_E_ 23. Cholycystokinin (CCK): *Excitatory*
A. stimulates gastric peristalsis
B. inhibits release of bile
C. stimulates release of pancreatic hormones
D. inhibits intestinal motility
E. stimulates release of trypsinogen and chymotrypsinogen

_A_ 24. This hormone secreted by the small intestinal mucosa stimulates the intercalated duct cells of the pancreas to secrete sodium bicarbonate.
A. secretin *Duodenum*
B. cholecystokinin
C. gastrin
D. enterokinase
E. enterogastrin

_D_ 25. The organ of greatest importance in providing the enzymes required for the digestion of food is the:
A. parotid gland   D. pancreas
B. stomach   E. liver
C. liver

OUTLINE 24

LEARNING OBJECTIVES (√)

After reading the assigned pages in the textbook, and/or reading and performing related laboratory exercises, and listening to lecture and/or laboratory presentations, you should be able to:

☐1. Diagram the structures of adenosine triphosphate (ATP) and adenosine diphosphate (ADP) and write the reactions for ATP dephosphorylation and ADP phosphorylation.

☐2. Explain the formation of ATP during the metabolism of one mole of glucose to CO_2, H_2O, and E. Include summary reactions of Anaerobic Glycolysis, Conversion of Pyruvate to Acetyl-CoA, Tricarboxylic Acid Cycle, and Oxidative Phosphorylation.

☐3. Explain how glycerol, amino acids, and fatty acids can be metabolized for the purpose of generating ATP.

☐4. Define Basal Metabolic Rate (BMR) and give the range of normal adult values.

☐5. Describe the effects of the following factors on metabolic rate: sex, body surface area, core body temperature, age, and secretion of calorigenic hormones.

☐6. Explain how metabolic rate can be indirectly determined by measuring the oxygen consumption of the subject and applying the caloric value of one liter of oxygen.

☐7. Define poikilothermy and homeothermy and give a range of normal values for human core body temperature.

☐8. Explain how each of the following mechanisms can result in a gain of body heat: basal metabolism, shivering, secretion of calorigenic hormones, and the Q_{10} effect of temperature.

☐9. Explain how each of the following mechanisms can result in a loss of body heat: radiation, evaporation, conduction, and convection.

☐10. Draw a negative feedback control system diagram to explain how the hypothalamus normally responds to a fall below normal of core body temperature. Include four physiological mechanisms of heat gain /heat conservation.

☐11. Draw a negative feedback control system diagram to explain how the hypothalamus normally responds to an increase above normal of core body temperature. Include four physiological mechanisms of heat loss and/or decreased heat production.

☐12. Define fever, pyrogen and antipyretic.

☐13. Explain why the hypothalamus maintains the febrile state until the fever breaks.

☐14. Define cold death and explain why it occurs.

☐15. Define heat death and explain why it occurs.

NOTES

DATE : _____

OUTLINE #_____

QUESTIONS

1._____
2._____
3._____
4._____
5._____

OUTLINE 24

LECTURER:_____   NOTES_____   DATE:_____

I. **METABOLISM**
    A. Energy Transformation
        1. Adenosine triphosphate: structure

        2. ATP dephosphorylation, ADP phosphorylation
            Reactions:

        3. Formation of ATP
            a. Anaerobic glycolysis
            b. Conversion of pyruvate to acetyl Co - A
            c. Tricarboxylic acid cycle
            d. Oxidative phosphorylation

        4. Metabolism of glycerol, fatty acid, amino acid

        5. Phosphocreatine
            a. Structure
            b. Dephosphorylation and E transfer to ATP

    B. Metabolic Rate
        1. Definition and normal range

        2. Factors affecting metabolic rate
            a. Sex

            b. Body surface area

            c. Core body temperature

            d. Age

            e. Calorigenic hormones

NOTES

DATE : _____

OUTLINE #_____

QUESTIONS

1._____
2._____
3._____
4._____
5._____

ENERGY TRANSFORMATION

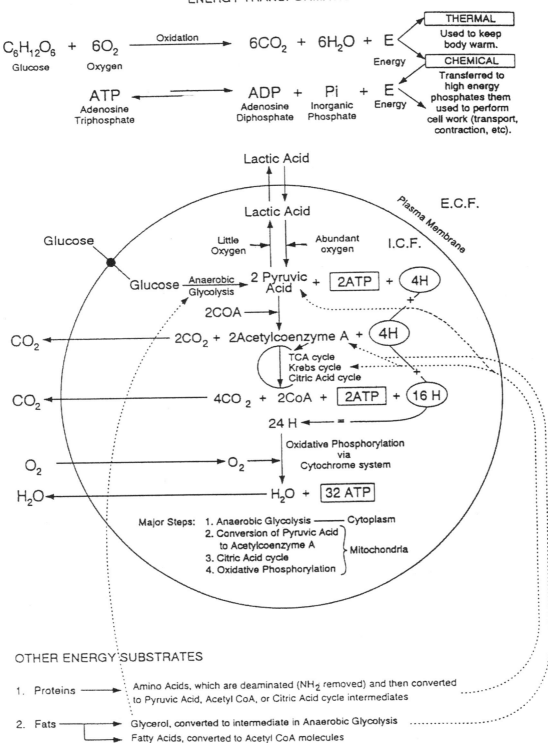

$$C_6H_{12}O_6 + 6O_2 \xrightarrow{\text{Oxidation}} 6CO_2 + 6H_2O + E$$

Glucose   Oxygen

THERMAL
Used to keep body warm.

Energy

CHEMICAL
Transferred to high energy phosphates them used to perform cell work (transport, contraction, etc).

$$ATP \longleftarrow \longrightarrow ADP + Pi + E$$

Adenosine Triphosphate   Adenosine Diphosphate   Inorganic Phosphate   Energy

Lactic Acid

Lactic Acid

E.C.F.

Glucose

Plasma Membrane

Little Oxygen   Abundant oxygen

I.C.F.

Glucose $\xrightarrow[\text{Glycolysis}]{\text{Anaerobic}}$ 2 Pyruvic Acid + 2ATP + 4H

2COA →

CO_2 ← $2CO_2$ + 2Acetylcoenzyme A + 4H

TCA cycle
Krebs cycle
Citric Acid cycle

CO_2 ← $4CO_2$ + 2CoA + 2ATP + 16 H

24 H ←

Oxidative Phosphorylation via Cytochrome system

O_2 → O_2 →

H_2O ← H_2O + 32 ATP

Major Steps: 1. Anaerobic Glycolysis —— Cytoplasm
2. Conversion of Pyruvic Acid to Acetylcoenzyme A
3. Citric Acid cycle
4. Oxidative Phosphorylation   } Mitochondria

OTHER ENERGY SUBSTRATES

1. Proteins ⟶ Amino Acids, which are deaminated (NH_2 removed) and then converted to Pyruvic Acid, Acetyl CoA, or Citric Acid cycle intermediates

2. Fats ⟶ Glycerol, converted to intermediate in Anaerobic Glycolysis
⟶ Fatty Acids, converted to Acetyl CoA molecules

NOTES

DATE : _____   *OUTLINE #_____*

QUESTIONS

1._____

2._____

3._____

4._____

5._____

3. Indirect measurement of BMR
   a. Basal conditions

   b. Measurement of O_2 consumption/Hr.

   c. Caloric value of oxygen

   d. Ht. & Wt. and body surface area

   e. Computation:

II. TEMPERATURE REGULATION
A. Poikilothermy vs. Homeothermy
   1. Poikilotherm

   2. Homeotherm

B. Body Heat Gain
   1. Basal metabolism

   2. Shivering

   3. Calorigenic hormones

   4. Q_{10} effect

C. Body Heat Loss
   1. Radiation

   2. Evaporation

   3. Conduction

   4. Convection

NOTES

DATE : _____   *OUTLINE #_____*

QUESTIONS

1._____
2._____
3._____
4._____
5._____

D. Thermoregulatory Mechanisms
    1.   Hypothalamic set - point

    2.   Hypothalamic heat gain center
        a.   Peripheral vasoconstriction

        b.   Shivering

        c.   Secretion of calorigenic hormones

        d.   Piloerection

    3.   Hypothalamic heat loss center
        a.   Peripheral vasodilation

        b.   Sweating

        c.   Panting

        d.   Lethargy

E.   Fever: Resetting Hypothalamic Set-Point

F.   Heat Death and Cold Death
    1.   Heat death

    2.   Cold death

NOTES

DATE : _____   *OUTLINE #_____*

QUESTIONS

1. _____
2. _____
3. _____
4. _____
5. _____

REVIEW QUESTIONS

I.   Matching

A.   Convection
B.   Conduction
C.   Radiation
D.   Evaporation
E.   Piloerection

_____1.   loss or gain of heat by air currents over the surface of the body
_____2.   quantitatively the most significant means of losing excess heat when the humidity is low and the environmental temperature is near or above surface body temperature
_____3.   the only non-evaporative means of heat loss not requiring air currents or body contact
_____4.   a mechanism to trap air near the body's surface for insulation against loss of heat
_____5.   the physical transfer of heat between two objects of different temperature in contact with each other

II.   Multiple choice

_____6.   Which of the following is not a homeotherm?
   A.   dog
   B.   cat
   C.   snake
   D.   horse
   E.   two of the preceding

_____7.   Walking fast (5 MPH) results in an energy expenditure of about 650 calories per hour. Assuming all of the energy required came from burning fat, how far would a person have to walk fast to burn the fat in one slice of cheese (about 10 grams of fat)?
   A.   90 feet
   B.   0.6 mile
   C.   about a mile and a half
   D.   130 feet
   E.   a quarter mile

_____8.   A naked person standing in a room with an air temperature of 72°F, no circulation of air and no sunlight, will lose body heat primarily by:
   A.   convection
   B.   evaporation
   C.   conduction
   D.   radiation
   E.   insensible perspiration

_____9.   Most of the ATP generated during the metabolism of one mole of glucose to CO_2, H_2O, and E is generated in the process of:
A.   anaerobic glycolysis
B.   conversion of pyruvate to acetyl COA
C.   the Kreb's Cycle (Tricarboxylic Acid Cycle)
D.   oxidative phosphorylation
E.   conversion of pyruvate to lactate

_____10.   The energy liberated by catabolic processes can be converted into:
A.   metabolic heat
B.   external work
C.   chemical storage
D.   both A and B
E.   A, B, and C

_____11.   An antipyretic:
A.   acts on the hypothalamus to cause peripheral vasoconstriction
B.   stimulates the hypothalamic heat - gain center
C.   lowers the hypothalamic set-point for core body temperature
D.   stimulates the release of TSH
E.   does none of the above

_____12.   After exercise, humans continue to use oxygen at greater than rest levels. The difference between post-exercise and resting levels of O_2 uptake is known as:
A.   anaerobic glycolysis
B.   formation of lactate from pyruvate
C.   oxidative phosphorylation
D.   the citric acid cycle
E.   the oxygen debt

_____13.   Given the following data, compute BMR:
caloric equivalent of 1 liter of oxygen = 4.825.
body surface area = 1.5 square meters.
6 minute oxygen consumption (Corrected to STP) = 1500 ml.
The answer is: _____ Cal/M2/Hr (rounded off).
A.   38
B.   48
C.   40
D.   42
E.   72

_____14.   How many calories per hour does the average adult's metabolic rate burn?
A.   300
B.   40
C.   70
D.   120
E.   200

_____15. Quantitatively the most significant process by which a human loses heat when the environmental temperature exceeds the body temperature is:
  A.   convection
  B.   evaporation
  C.   radiation
  D.   piloerection
  E.   conduction

_____16. Which of the following temperature regulating mechanisms is not controlled by the hypothalamic heat-gain center?
  A.   shivering
  B.   peripheral vasoconstriction
  C.   increased BMR
  D.   sweating
  E.   T3 secretion

_____17. A 70 kilogram person at rest is consuming 0.25 liter of oxygen per minute. What is the metabolic rate in calories per hour?
  A.   42.1
  B.   15.0
  C.   17.5
  D.   38
  E.   72.4

_____18. Which of the following is NOT characteristic of conditions prevailing for a BMR test?
  A.   fasting for at least 12 hours
  B.   person at complete rest for 30 to 60 minutes prior to test
  C.   patient received an enema one to two hours before test
  D.   surrounding temperature approximately 25 degrees centigrade
  E.   usually in the morning after a restful sleep

_____19. Assuming 75°F environmental temperature on a windless, cloudy day, the principal means of non-evaporative heat loss by a person dressed in shorts and a T-shirt and walking is:
  A.   radiation
  B.   conduction
  C.   convection
  D.   sweating
  E.   piloerection

_____20. During anaerobic glycolysis:
  A.   oxygen is consumed and carbon dioxide is produced
  B.   4 moles of ATP are formed per mole of glucose metabolized
  C.   4 hydrogen atoms are produced
  D.   two of the above
  E.   none of the above

_____21.   Pyretics:
  A.   act on the hypothalamus to cause peripheral vasodilatation
  B.   stimulate the hypothalamic heat-gain center
  C.   lower the hypothalamic set-point for core body temperature
  D.   are chemicals that reduce fever
  E.   inhibit the release of TSH

_____22.   The process of oxidative phosphorylation:
  A.   occurs within cytoplasm
  B.   generates hydrogen atoms
  C.   produces water
  D.   consumes 24 moles of ATP
  E.   two of the preceding

_____23.   When ambient air temperature rises above skin surface temperature, the
             following heat-loss factor becomes a heat-gain factor:
  A.   evaporation
  B.   radiation
  C.   convection
  D.   conduction (object cooler)
  E.   none of the above

_____24.   An animal that regulates its body core temperature within a narrow range of
             acceptable values is called:
  A.   a geotherm   D.   a poikilotherm
  B.   a homeotherm   E.   a thermal
  C.   an isotherm

OUTLINE 25

LEARNING OBJECTIVES (√)

After reading the assigned pages in the textbook, and/or reading and performing related laboratory exercises, and listening to lecture and/or laboratory presentations, you should be able to:

☐1. Define each of the following hormonal control mechanisms: autocrine, paracrine, endocrine, and neuroendocrine.

☐2. Define the term second messenger as it applies to hormone activity at the cellular level.

☐3. Diagram the following second messenger systems and explain where applicable the functions of G-proteins, and give an example of each: cyclic AMP, diacylglycerol-IP$_3$, and receptor-linked ion channels.

☐4. Diagram the gene expression model of hormone action and give an example.

☐5. Explain the function of atrial natriuretic peptide (ANP), giving its source, stimulus for release, and target organ, and activity.

☐6. Describe the location of the pineal gland, list the sources and physiological activities of melatonin, and outline mechanisms that control melatonin secretion.

☐7. Diagram the general structure of the hypophysis (pituitary gland) and its relation to the hypothalamus.

☐8. List the general functions of human growth hormone (GH) or (somatotropin, STH), describe factors that influence its rate of secretion, and define pituitary giantism, acromegaly, and pituitary dwarfism.

☐9. Outline the functions of thyroid stimulating hormone (TSH), and adrenocorticotropic hormone (ACTH).

☐10. Define gonadotropin, and briefly describe the functions of follicle stimulating hormone (FSH) , luteinizing hormone (LH) (ICSH), and prolactin.

☐11. Briefly outline the source and function of melanocyte stimulating hormone (MSH) (intermedin).

☐12. Describe the physiological activity of antidiuretic hormone and explain its connection to diabetes insipidus.

☐13. Describe the function of oxytocin.

☐14. Draw a diagram of the hypothalamic-hypophyseal portal system showing the functional connection between the hypothalamus and the anterior and middle pituitary.

☐15. List, where known, the hypothalamic releasing hormone and/or the release- inhibiting hormone for each of the following hormones: STH, TSH, ACTH, FSH, LH, Prolactin, MSH.

NOTES

DATE : _____   *OUTLINE #_____*

QUESTIONS

1._____
2._____
3._____
4._____
5._____

OUTLINE 25

LECTURER:_____ NOTES_____ DATE:_____

I. **ENDOCRINE REGULATION**
   A. Comparative Control Mechanisms
   Diagram: comparison of endocrine control mechanisms
   1. Autocrine control

   2. Paracrine control

   3. Endocrine control

   4. Neuroendocrine control

   B. Modes of Hormone Action
   Diagrams: second messenger systems
   1. Second messenger systems
      a. Cyclic AMP

      b. Diacylglycerol- IP3

      c. Receptor-linked ion channels

   2. Gene expression model

II. **THE HEART**
   Diagram: ANP regulation of blood pressure
   A. Atrial Natriuretic Peptide (ANP)
   1. Source

   2. Stimulus for release

   B. Blood Pressure and Blood Volume Regulation

NOTES

QUESTIONS

1._____
2._____
3._____
4._____
5._____

III. THE PINEAL GLAND
   A. Location

   B. Melatonin
      1. Sources

      2. Physiological activities

      3. Control of melatonin secretion

IV. THE HYPOPHYSIS (PITUITARY GLAND)
   Diagram: the pituitary gland and the hypothalamus
   A. General Structure and Location
      1. Hypothalamus
      2. Anterior pituitary (adenohypophysis)
      3. Middle pituitary (pars intermedia)
      4. Posterior pituitary (neurohypophysis)

   B. Anterior Lobe Hormones
      1. Growth hormone (GH, STH, Somatotropin)
         a. Functions

         b. Giantism

         c. Acromegaly

         d. Dwarfism

      2. Thyroid stimulating hormone (Thyrotropin, TSH

      3. Adrenocorticotropic hormone (ACTH, Corticotropin)

NOTES

DATE : _____   *OUTLINE #_____*

QUESTIONS

1._____
2._____
3._____
4._____
5._____

4.   Follicle stimulating hormone (FSH)

5.   Luteinizing hormone (LH, ICSH)

6.   Prolactin (luteotropic hormone, LTH, lactogenic hormone)

C.   Middle Lobe Hormone: Melanocyte Stimulating Hormone (MSH, Intermedin)

D.   Posterior Lobe Hormones
        1.   Antidiuretic hormone
                a.   Physiological activity

                b.   Diabetes insipidus

        2.   Oxytocin

V.   THE HYPOTHALAMUS

A.   The Hypothalamic-Hypophyseal Portal Circulation
        Diagram: arterial supply, primary capillary plexus, portal vessels, secondary capillary plexus, venous drainage

NOTES

QUESTIONS

1._____
2._____
3._____
4._____
5._____

B. Releasing Hormones and Release-inhibiting Hormones for:

Pituitary Hormones	Hypothalamic Releasing Hormone	Hypothalamic Release-Inhibiting Hormone
STH		
TSH		
ACTH		
FSH		
LH		
Prolactin		
MSH		

C. The H-H portal system and control of core body temperature.
Diagram: negative feedback control of core temperature

NOTES

DATE : _____   *OUTLINE #_____*

QUESTIONS

1._____
2._____
3._____
4._____
5._____

REVIEW QUESTIONS

I. Matching

A. L.H
B. T.S.H
C. A.C.T.H
D. S.T.H
E. Prolactin

_____1. hypersecretion may result in goiter
_____2. stimulates secretion of breast milk
_____3. causes maturation of ovarian follicle
_____4. directly elevates plasma glucocorticoids
_____5. hyposecretion may result in pituitary dwarfism

II. Multiple choice

_____6. The second messenger system may involve:
A. adenylate cyclase/cyclic AMP
B. phosphatidylinositol and diacylglycerol
C. receptor linked ion channels
D. both (A) and (B)
E. (A), (B), and (C)

_____7. Which of the following is associated with an increase in the intracellular formation of cyclic AMP?
A. B- protein
B. D- protein
C. Gi protein
D. Gs protein
E. myosin

_____8. Which of the following stimulates the release of ADH?
A. high ECF osmotic pressure
B. ANF
C. T_4
D. urea
E. two of the preceding

_____9. Hypersecretion of HGH (also known as STH) after adulthood has been reached results in:
A. cretinism
B. acromegaly
C. Addison's disease
D. Cushing's disease
E. diabetes insipidus

_____10.   TSH - RH:
A.   release is accelerated when ECF osmotic pressure decreases below normal
B.   indirectly promotes elevation of plasma T_3
C.   stimulates cells of the neurophypophysis
D.   was the first hypothalamic inhibitory hormone to be discovered
E.   release is depressed when plasma iodine levels decrease below normal

_____11.   Atrial natriuretic peptide (ANP):
A.   increases systemic arterial blood pressure
B.   is an endopeptidase
C.   increases water permeability of the distal nephron
D.   inhibits the release of ADH
E.   increases sodium reabsorption

_____12.   Light falling on the retina inhibits the release of this hormone. The hormone is:
A.   melanocyte stimulating hormone
B.   norepinephrine
C.   melatonin
D.   parathormone
E.   insulin

_____13.   In the _____ mode of intercellular signaling a secretory cell releases a chemical messenger which then engages a receptor on another part of the same cell.
A.   neuroendocrine
B.   holocrine
C.   exocrine
D.   autocrine
E.   paracrine

_____14.   Intermedin:
A.   is also called ICSH
B.   stimulates testosterone secretion
C.   controls dispersion of melanin in the skin
D.   is required for maturation of the ovarian follicle
E.   causes metabolic rate to double

_____15.   Which of the following is a principal stimulator of androgen secretion by the testis?
A.   prolactin
B.   oxytocin
C.   interstitial Cell Stimulating Hormone
D.   TSH-RH
E.   ACTH

359

_____16. An inhibitory hormone that engages a cell surface receptor may exert its effects on the target cell by way of:
  A.   increasing cyclic AMP formation
  B.   an inhibitory G-protein
  C.   stimulating Ca^{++} release
  D.   turning on one or more of the cell's genes
  E.   blocking a sodium channel

_____17. In comparing endocrine control and neural control, it is true to say that in general:
  A.   endocrine controls are much faster
  B.   neural controls are longer-lasting
  C.   neural controls are more specific or localized
  D.   the two systems have opposite effects on most target cells
  E.   most organs are controlled by one or the other of the two control systems but not both

_____18. STH - RH (HGH-RH) is:
  A.   a gonadotropin
  B.   an anterior pituitary hormone controlling ovarian function
  C.   an abbreviation for growth hormone
  D.   a hypothalamic hormone controlling release of growth hormone
  E.   a hormone that regulates ECF osmotic pressure

_____19. Membrane hormone receptors may be associated with:
  A.   B-proteins   D.   P-proteins
  B.   D-proteins   E.   R-proteins
  C.   G-proteins

_____20. The element used as an intracellular signal in hormone mediated responses in target cells is:
  A.   calcium   D.   magnesium
  B.   chlorine   E.   potassium
  C.   iron

_____21. An excitatory hormone that engages a cell surface receptor may exert its effects on the target cell by way of
  A.   decreasing cyclic AMP formation
  B.   an inhibitory G-protein
  C.   inhibiting Ca^{++} release
  D.   turning off one or more of the cell's genes
  E.   opening a calcium channel

_____22. In a paracrine control mechanism, the secreted hormone usually:
A. travels via the blood to distant target cells
B. engages a receptor on the originating cell surface
C. the nervous system and endocrine system work together to exert control of the target cells
D. engages receptors on nearby target cells
E. enters the cell and engages a receptor in the nucleus

_____23. Antidiuretic hormone release from the posterior pituitary is stimulated by:
A. a fall in plasma osmolality
B. severe hemorrhage
C. stimulation of arterial barorecptors
D. stretch of left atrial receptors
E. two of the above

_____24. The primary stimulus for _____secretion in a female is suckling by a nursing infant.
A. FSH
B. LH
C. TSH
D. oxytocin
E. ADH

OUTLINE 26
LEARNING OBJECTIVES (√)

After reading the assigned pages in the textbook, and/or reading and performing related laboratory exercises, and listening to lecture and/or laboratory presentations, you should be able to:

☐1. Describe the location and the general structure of the thyroid gland, and with the aid of a diagram, identify the lobes , isthmus, follicular cells, parafollicular cells, and colloid.

☐2. Diagram the synthesis, storage, secretion, and transport of T_3 (triiodothyronine) and T_4 (tetraiodothyronine).

☐3. Outline the general functions of T_3 and T_4 and briefly list the general systemic effects of hyposecretion and hypersecretion.

☐4. Diagram the negative feedback system that controls T_3 and T_4 secretion. Include the following elements: hypothalamus, anterior pituitary, TRH,TSH, follicular cell, and plasma T_3 and T_4.

☐5. Define goiter and explain how a goiter can develop in either a hypothyroid state, a hyperthyroid state, or a euthyroid state.

☐6. Explain the cause and effects of each of the following disorders associated with the thyroid gland: myxedema, cretinism, exophthalmus, Grave's disease, and Hashimoto's disease.

☐7. Describe the general structure and cell types of the endocrine pancreas and list the cells' endocrine secretions.

☐8. Outline the physiological activities of each of the following pancreatic hormones: insulin, glucagon, somatostatin, and pancreatic polypeptide.

☐9. Draw a negative feedback control system diagram to show how insulin and glucagon regulate blood glucose. Include the normal range of blood glucose concentrations.

☐10. Explain the systemic effects of unmanaged diabetes mellitus.

☐11. Explain the differences between type I diabetes mellitus (IDDM) and type II diabetes mellitus (NIDDM) relative to cause, age of onset, symptoms, and management.

☐12. Describe the location and general structure of the adrenal glands. Include a description of the adrenal medulla and the following zones of the adrenal cortex: glomerulosa, fasciculata, and reticularis.

☐13. Define mineralocorticoid, give an example, outline its physiological activities, and describe three modes of control of secretion.

☐14. Define glucocorticoid, give an example, outline its physiological activities, and diagram a negative feedback control of secretion.

☐15. Describe Addison's disease and Cushing's disease.

☐16. Describe the location and general structure of the parathyroids.

☐17. Explain how PTH and calcitonin control ECF $[Ca^{++}]$.

NOTES

DATE : _____   *OUTLINE #_____*

QUESTIONS

1._____
2._____
3._____
4._____
5._____

OUTLINE 26

LECTURER:_____   NOTES_____   DATE:_____

I.   **THE THYROID GLAND**
     Diagram: location and structure of thyroid gland
     A.   General Structure and Location
           1.   Lobes
           2.   Isthmus
           3.   Follicular cells
           4.   Parafollicular cells
           5.   Colloid

     B.   Tetraiodothyronine (T_4) and Triiodothyronine (T_3)
           Diagram: synthesis, release, transport of T_3 and T_4
           1.   Synthesis, release, transport

           2.   Hypersecretion and hyposecretion
                a.   Goiter

                b.   Myxedema

                c.   Cretinism

                d.   Exopthalmus

                e.   Grave's disease

                f.   Hashimoto's disease

NOTES

QUESTIONS

1._____

2._____

3._____

4._____

5._____

II. THE ENDOCRINE PANCREAS: ISLETS OF LANGERHANS
Diagram: location and structure of pancreas

A. General Structure and Cell Types
1. Alpha cell
2. Beta cell
3. Delta cell
4. F cell

B. Hormone Activities
1. Insulin

2. Glucagon

3. Somatostatin

4. Pancreatic polypeptide

C. Control of Blood Glucose: Insulin and Glucagon
Diagram: negative feedback control of blood sugar

D. Diabetes Mellitus
1. Systemic effects

2. Type I: *Insulin - Dependent Diabetes Mellitus (IDDM)*

3. Type II: *Non-Insulin-Dependent Diabetes Mellitus (NIDDM)*

NOTES

DATE : _____   *OUTLINE #_____*

QUESTIONS

1._____
2._____
3._____
4._____
5._____

III. THE ADRENAL GLANDS
Diagram: adrenal gland location and internal structure

- **A.** General Location and Structure
  1. Cortex
     - a. Zona glomerulosa
     - b. Zona fasciculata
     - c. Zona reticularis
  2. Medulla

- **B.** Mineralocorticoids
  1. Activities

  2. Control of secretion
     - a. Renin-angiotensin system

     - b. Hypothalamic-hypophyseal system

     - c. Direct effect of ECF Na^+/K^+

- **C.** Glucocorticoids
  1. Activities

  2. Control of secretion
     Diagram: negative feedback control: stress vs. cortisol secretion

NOTES

DATE : _____   *OUTLINE #*_____

QUESTIONS

1._____
2._____
3._____
4._____
5._____

D.   Androgens and Estrogens

E.   Hyposecretion of the Adrenal Cortex: *Addison's Disease*

F.   Hypersecretion of the Adrenal Cortex: *Cushing's Disease*

IV.   **THE PARATHYROID GLANDS**
      Diagram: location and internal structure of parathyroids
      A.   General Structure and Location

      B.   Parathyroid Hormone
             1.   Physiological activities

             2.   Control of secretion

V.   **ECF Ca^{++} REGULATION: PTH AND CALCITONIN**
      Diagram: negative feedback control of ECF Ca^{++}

NOTES

QUESTIONS

1._____
2._____
3._____
4._____
5._____

REVIEW QUESTIONS

I.   Matching

       A.   Calcitonin
       B.   Glucagon
       C.   P.T.H.
       D.   Insulin
       E. T_3

_____1.   a thyroid hormone secreted by parafollicular cells
_____2.   a hyperglycemic hormone secreted by the pancreas
_____3.   secreted by follicular cells; contains iodine
_____4.   a hypoglycemic hormone; secreted by the pancreas
_____5.   stimulates osteoclasts; raises ECF Ca^{++}

II.   Multiple choice

_____6.   Glucagon:
       A.   is a glucocorticoid
       B.   stimulates hepatic glycogenolysis
       C.   functions as an antagonist to hydrocortisone
       D.   inhibits gluconeogenesis
       E.   two of the preceding.

_____7.   Hyposecretion of _____ may lead to tetany and death.
       A.   calcitonin
       B.   insulin
       C.   PTH
       D.   LTH
       E.   ACTH

_____8.   Hypersecretion of _____, if severe, may cause calcium phosphate to precipitate in tissues such as lung and heart causing death.
       A.   calcitonin
       B.   insulin
       C.   PTH
       D.   LTH
       E.   ACTH

_____9.   In a normal adult oral glucose tolerance test, blood sugar "peaks" about _____ after ingestion.
       A.   10-20 minutes
       B.   30-60 minutes
       C.   60-90 minutes
       D.   20-120 minutes
       E.   120-180 minutes

_____10.   Parathyroid hormone:
   A.   decreases ECF calcium levels
   B.   stimulates osteoblasts
   C.   is secreted by perifollicular cells
   D.   stimulates osteoclasts
   E.   two of the preceding

_____11.   Triiodothyronine production and release is controlled directly by the:
   A.   hypothalamus
   B.   neurohypophysis
   C.   adenohypophysis
   D.   middle pituitary
   E.   parathyroids

_____12.   Insulin:
   A.   promotes cellular uptake of glucose
   B.   controls ECF calcium levels
   C.   is a hyperglycemic hormone
   D.   is secreted by pancreatic alpha cells
   E.   two of the preceding

_____13.   Thyroxin contains:
   A.   steroid
   B.   threonine
   C.   tyrosine
   D.   TSH
   E.   none of the above

_____14.   Type II diabetes mellitus:
   A.   is due to hyposecretion of glucagon
   B.   is always treated by insulin administration
   C.   is due to hypersecretion of insulin
   D.   involves a lack of insulin affect on target cells due to a change in insulin receptor function
   E.   two of the above

_____15.   When plasma calcium rises above normal, _____ secretion is stimulated:
   A.   aldosterone
   B.   calcitonin
   C.   norepinephrine
   D.   ADH
   E.   parathyroid hormone

_____16.   Calcium is required for all but which one of the following?
     A.   maintenance of normal sodium permeability in nerves
     B.   blood clotting to occur
     C.   formation of visual pigment
     D.   secretion of certain proteins
     E.   serving as a second messenger to allow some hormones to operate

_____17.   A decrease in serum calcium promotes, for corrective purposes:
     A.   release of calcitonin
     B.   hypersecretion of TRH
     C.   release of insulin
     D.   release of PTH
     E.   release of cortisone

_____18.   A fall in core body temperature below the hypothalamic set-point may result in an increased:
     A.   thyroid release of T_3
     B.   hypothalamic release of oxytocin
     C.   adenohypophyseal release of SRH
     D.   adenohypophyseal release of FSH
     E.   secretion of enterogastrone

_____19.   Glucagon:
     A.   is a glucocorticoid
     B.   stimulates hepatic glycogenolysis
     C.   functions as an antagonist to hydrocortisone
     D.   inhibits gluconeogenesis
     E.   two of the preceding

_____20.   Cushing's disease is characterized by:
     A.   excessive retention of salt and water
     B.   decreased resistances to stress because of inadequate glucocorticoid secretion
     C.   lowered blood volume, blood pressure, and cardiac output
     D.   excessive retention of potassium
     E.   none of the preceding

_____21.   Addison's disease is characterized by:
     A.   excessive retention of salt and water
     B.   elevated blood volume, blood pressure, and cardiac output
     C.   increased levels of adrenocortical sex hormones
     D.   depletion of extracellular sodium and water
     E.   adrenal diabetes

_____22.   Adrenal diabetes:
    A.   may occur as a complication of Cushing's disease
    B.   can result from hyposecretion of the adrenal cortex
    C.   involves renal loss of water but not sugar
    D.   is more commonly known as Addison's disease
    E.   is treated by administration of cortisone

_____23.   Calcium and phosphate serum levels are controlled by:
    A.   parathryoid hormone
    B.   thyroid hormones T_3 and T_4
    C.   glucagon
    D.   insulin
    E.   cholecystokinin

OUTLINE 27

LEARNING OBJECTIVES (√)

After reading the assigned pages in the textbook, and/or reading and performing related laboratory exercises, and listening to lecture and/or laboratory presentations, you should be able to:

☐1. Diagram the basic structure of the male sex apparatus. Include the following components: testis, epidydymis, vas deferens, seminal vesicle, bulbourethral gland, prostate gland, urethra, penis, scrotum.

☐2. Define each of the following cells and their role in spermatogenesis: spermatogonium, primary spermatocyte, secondary spermatocyte, spermatid, spermatozoa, and Sertoli cells.

☐3. Distinguish between mitosis and meiosis and explain when each occurs during spermatogenesis.

☐4. List several functions of Sertoli cells.

☐5. Briefly describe the maturation and storage of sperm, and the composition and function of semen.

☐6. Draw a diagram of a negative feedback system that controls spermatogenesis and the development and maintenance of male secondary sex characteristics. Include the hypothalamus, gonadotropin-releasing hormone, anterior pituitary, FSH, LH, interstitial cells, Sertoli cells, testosterone, and inhibin.

☐7. List six testosterone-dependent changes at puberty.

☐8. Define sterility and impotence, and list two causes of each.

☐9. Diagram the basic structure of the female sex apparatus. Include the following components: ovary, uterine tube with fimbria, uterus, vagina, and external genitalia.

☐10. Define each of the following and their role in oogenesis and follicular development: oogonia, primary oocytes, primary follicle, zona pellucida, granulosa cells, theca interna, and ovum.

☐11. Define the follicular phase, the luteal phase, and the menstrual phase of the ovarian cycle, and indicate the time at which ovulation typically occurs.

☐12. Draw a negative feedback control system diagram to explain neuroendocrine control of the ovarian cycle.

☐13. Briefly list the physiological effects of estrogens and progesterone.

☐14. Outline the processes of fertilization and implantation.

☐15. List five functions of the placenta

☐16. Briefly describe the germinal stage, the embryonic stage, and the fetal stage of fetal development, including the duration of each stage.

☐17. Define parturition and outline its hormonal control.

☐18. Describe three methods of birth control and their effectiveness.

NOTES

DATE : _____   *OUTLINE #_____*

QUESTIONS

1._____
2._____
3._____
4._____
5._____

OUTLINE 27

LECTURER:_____   NOTES_____   DATE:_____

I.   **REPRODUCTIVE ENDOCRINOLOGY: MALE**
     Diagram: sagittal section; reproductive system
     A.   Basic Structure of Male Sex Apparatus
          1.   Testis
          2.   Epididymis
          3.   Vas deferens
          4.   Seminal vesicle
          5.   Bulbourethral gland
          6.   Prostate gland
          7.   Ejaculatory duct
          8.   Urethra

     B.   Spermatogenesis
          Diagram: seminiferous tubules and formation of spermatocytes

     C.   Composition and Function of Semen

     D.   Hormonal Control of Secondary Sex Characteristics

     E.   Sterility and Impotence

NOTES

DATE : _____   OUTLINE #_____

QUESTIONS

1._____
2._____
3._____
4._____
5._____

II. REPRODUCTIVE ENDOCRINOLOGY: FEMALE

Diagram: coronal section; reproductive system

A. Basic Structure of Female Sex Apparatus
1. Ovary
2. Uterine tube with fimbria
3. Uterus
4. Vagina

B. Oogenesis and the Menstrual Cycle
Diagram: phases of the menstrual cycle and maturation of the ovum

C. Ovulation and Fertilization

D. Pregnancy and Lactation

E. Birth Control

F. Sterility

NOTES

DATE : _____   *OUTLINE #*_____

QUESTIONS

1._____
2._____
3._____
4._____
5._____

REVIEW QUESTIONS

I. Matching

A. Testes
B. Epididymis
C. Vas deferens
D. Seminal vesicles
E. Prostate

_____1. produces an alkaline fluid that helps maintain viability of sperm
_____2. sperm storage
_____3. sperm production: mitosis, meiosis, differentiation
_____4. sperm transport and maturation – motility and fertility
_____5. production of seminal fluid containing nutrients, fructose, and prostaglandins

II. Multiple choice

_____6. Fertilization of the ovum normally occurs in the:
A. vagina
B. uterus
C. abdominal cavity
D. uterine tube
E. cervix

_____7. Which of the following is NOT an action of the placenta?
A. delivery of waste from fetus to mother
B. exchange of gases between fetus and mother
C. delivery of antibodies from mother to fetus
D. delivery of water and electrolytes from mother to fetus
E. delivery of nutrients from fetus to mother

_____8. Two gonadotropins work together to produce mature ova. They are:
A. LTH and STH
B. LH and ACTH
C. TSH and FSH
D. FSH and LH
E. LH and LTH

_____9. The time between fertilization of the ovum and implantation of the fertilized ovum in the uterus is approximately:
A. less than 24 hours
B. 2 days
C. 5 to 10 days
D. 21 days
E. one month

_____10. Which hormone's plasma concentration peaks (is at its highest) about a week after ovulation?
   A.   estrogen
   B.   progesterone
   C.   LH
   D.   FSH
   E.   none of the preceding

_____11. During the menstrual cycle _____ inhibits the release of gonadotropin releasing hormone.
   A.   ACTH
   B.   estrogen
   C.   prolactin release inhibiting hormone
   D.   testosterone
   E.   renin

_____12. The blood concentration of _____ "peaks" last during the ovarian cycle of a normal 24 year old female:
   A.   estrogen
   B.   FSH
   C.   progesterone
   D.   LH
   E.   STH

_____13. The corpus luteum is maintained during early pregnancy by:
   A.   progesterone
   B.   follicle stimulating hormone
   C.   human chorionic gonadotropin
   D.   estrogen
   E.   prolactin

_____14. Testosterone:
   A.   enlarges the vocal cords
   B.   thickens the skin
   C.   stimulates muscular development
   D.   causes bone growth
   E.   all of the above

_____15. Midway in a 28-day menstrual cycle:
   A.   serum LH is at its peak (high concentration)
   B.   serum progesterone is at its peak
   C.   the ovary stops producing estrogens
   D.   the corpus albicans forms
   E.   two of the preceding

_____16. Estrogens (during the follicular phase of the ovarian cycle):
   A.   stimulate the hypothalamus
   B.   stimulate the anterior pituitary
   C.   stimulate the ovaries
   D.   all of the above

_____17.   The ovarian cycle consists of the:
        A.   menstrual phase
        B.   luteal phase
        C.   follicular phase
        D.   both A and C
        E.   A, B, and C

_____18.   One of the following is a gonadotropin that stimulates spermatogenesis.
Which one?
        A.   LTH
        B.   TSH
        C.   HGH
        D.   LH
        E.   FSH

NOTES

QUESTIONS

1._____
2._____
3._____
4._____
5._____

MULTIPLE CHOICE. SELECT THE BEST ANSWER.

_B_ 1. Which of the following best fits the literal definition of "physiology"?
A. the study of stars
B. the study of nature
C. the study of animals
D. the study of plants
E. the study of religion

_____ 2. Historically, processes of the body were believed to be driven by various forms of "pneuma". The principal proponent of pneumatology was:
A. Harvey.
B. Vesalius.
C. Herophilus
D. Erasistratus
E. Galen

_____ 3. Placing an erythrocyte (red blood cell) into this solution will result in the cell crenating. This solution is:
A. 0.9% NaCl
B. 0.5% NaCl
C. 9.0% NaCl
D. 0.7% NaCl
E. 0.1% NaCl

_____ 4. Sometimes called the father of modern physiology, he was the discoverer of the circulation of blood. He was:
A. Priestly   D. Harvey
B. Galen   E. Malphighi
C. DaVinci

386

_____5.   Carrier molecules involved in facilitative diffusion or active transport:
A.   are membrane-bound proteins
B.   transport solvent as well as solute
C.   may be carbohydrate, lipid, or protein
D.   enter the plasma membrane on one surface and leave it from the other surface
E.   always carry solutes out of the cell

_____6.   As observed in similar laboratory experiments:
A.   0.09% NaCl + RBC = hemolysis
B.   0.09% NaCl + RBC = crenation
C.   0.09% NaCl + RBC = no change in RBC size or shape
D.   1.00% NaCl + RBC = hemolysis
E.   0.09% NaCl + RBC = crenation

_____7.   The organelle where most of the cell's proteins are synthesized is the:
A.   nucleus
B.   lysosome
C.   rough ER
D.   smooth ER
E.   centrosome

_____8.   Which of the following is not one of the five steps in the scientific method?
A.   observation   D.   hypothesis
B.   conclusion   E.   experiment
C.   deduction

_____9.   Which of the following requires a supply of metabolic energy because the solute moves "uphill"?
A.   dialysis
B.   osmosis
C.   facilitative diffusion
D.   net diffusion
E.   primary active transport

_____10.   Simple diffusion (net diffusion without carrier assistance) differs from facilitated diffusion because it:
A.   is much slower
B.   can be blocked by competitive inhibitors
C.   reaches a maximum rate
D.   requires expenditure of energy
E.   is not affected by a change in concentration gradient

_____11. A process called _____ involves invagination of the plasma membrane.
- A. osmosis
- B. facilitated diffusion
- C. exocytosis
- D. endocytosis
- E. simple diffusion

_____12. A secretory vesicle leaves a cell by:
- A. net diffusion
- B. exocytosis
- C. endocytosis
- D. phagocytosis
- E. pinocytosis

_____13. Transmission at the neuromuscular junction is dependent upon:
- A. curare
- B. acetylcholine
- C. noradrenalin
- D. serotonin
- E. actin & myosin

_____14. In which of the following would the velocity of nerve impulse conduction be the least?
- A. large diameter unmyelinated fibers
- B. small diameter unmyelinated fibers
- C. large diameter myelinated fibers
- D. small diameter myelinated fibers
- E. none of the preceding because they conduct impulses at the same velocity

_____15. One of the following statements about dialysis is false. Which one?
- A. A membrane is required.
- B. A solute concentration gradient must exist.
- C. It is a kind of net diffusion.
- D. It is affected by change in temperature.
- E. A membrane carrier may be required.

_____16. The Na^+ - K^+ pump is an example of:
- A. symport
- B. facilitated diffusion
- C. antiport
- D. dialysis
- E. carrier competition

_____17. In a process of facilitated diffusion:
  A. a substrate is moved against its concentration gradient
  B. the carrier mechanism is directly coupled to an energy-yielding reaction
  C. membrane carriers assist the "downhill" movement of a solute
  D. solvents are transported from higher to lower concentration
  E. membrane-bound carriers may or may not be required

_____18. Which of the following would be thought of as the first step in the scientific method?
  A. hypothesis
  B. publication
  C. experiment
  D. observation
  E. conclusion

_____19. Transmission at the neuromuscular junction is dependent upon:
  A. ach receptors
  B. acetylcholine
  C. cholinesterase
  D. sodium ions
  E. all of the above

_____20. Symport refers to carrier transport of a solute:
  A. against an electrical gradient
  B. coupled to and in the same direction as Na^+ transport
  C. coupled to but in the opposite direction as Na^+ transport
  D. against an osmotic gradient
  E. through the nuclear envelope

_____21. Dialysis:
  A. can occur without a membrane
  B. requires a membrane-bound carrier
  C. is directly coupled to energy-yielding reactions
  D. is a process of reverse osmosis
  E. is the net diffusion of a solute through a selectively permeable membrane

_____22. In the laboratory, you set up a thistle tube osmometer. Water moved from the beaker and into the dialysis bag which you had filled with a sugar solution. Water moved by the process of:
  A. osmosis
  B. dialysis
  C. simple diffusion
  D. facilitative diffusion
  E. active transport

_____23. A facilitated transport system can become saturated because:
    A.   all binding sites are filled
    B.   the channel is filled with diffusing molecules
    C.   the channel becomes saturated
    D.   the gate becomes closed
    E.   all of the above

_____24. Which is the last of five steps in the scientific method?
    A.   hypothesis
    B.   conclusion
    C.   experiment
    D.   observation
    E.   publication

_____25. The type of molecule that has the best chance to penetrate a cell membrane by simple diffusion is:
    A.   a small charged molecule
    B.   a small highly lipid soluble molecule
    C.   a small polar molecule
    D.   a large highly lipid soluble molecule
    E.   a large nonpolar molecule

_____26. A red blood cell is placed in a 0.9% NaCl solution and intracellular volume does not change because:
    A.   the salt solution is hypotonic
    B.   the salt concentration inside the cell is 0.9%
    C.   the salt solution is hypertonic
    D.   the plasma membrane is impermeable to water
    E.   intracellular and extracellular fluids are iso-osmotic

_____27. The smaller the motor unit arrangement in skeletal muscle:
    A.   the less precise the degree of neural control of the muscle
    B.   the greater the strength of skeletal muscle
    C.   the larger the size of the muscle
    D.   the more precise the degree of neural control of the muscle
    E.   two of the preceding

_____28. The fused, sustained, contraction of component fibers within a skeletal muscle is know as:
    A.   contracture
    B.   tetanus
    C.   treppe
    D.   fatigue
    E.   tonus

_____29. A physiological definition of homeostasis is:
  A. everything stays the same
  B. all for one and one for all
  C. the maintenance of a relatively stable internal environment
  D. likes repel, opposites attract
  E. what goes up must come down

_____30. Saltatory conduction refers to:
  A. the changes in membrane permeability that occur with depolarization
  B. the jumping of an action potential from one node of Ranvier to the next
  C. non-myelinated axons only
  D. the impulse jumping from a nerve to a muscle
  E. the active transport of ions across the nerve membrane

_____31. The maintenance of a relative stable internal environment defines the term:
  A. physiology
  B. negative feedback
  C. homeostasis
  D. osmosis
  E. none of the preceding

_____32. Which of the statements regarding plasma membranes is false?
  A. The polar parts of the lipid components are located on the inner and outer surfaces of the plasma membrane.
  B. The nonpolar core of the plasma membrane forms a significant barrier to materials crossing the membrane.
  C. A number of proteins in the plasma membrane form pores or channels that allow some materials to pass through.
  D. Chemical residues associated with reception of information are found on the inner surface of the plasma membrane.
  E. Cholesterol is a component of plasma membranes.

_____33. The principal function of this organelle is to generate high energy phosphate compounds that provide usable energy for cell work. It is the:
  A. smooth ER
  B. golgi complex
  C. mitochondrion
  D. lysosome
  E. nucleolus

391

_____34.	Sarcoplasmic reticula:
A.	synthesize proteins
B.	are involved in production of ATP
C.	are Ca^{++} storage areas
D.	are K^+ storage areas
E.	open into the T tubules

_____35.	A literal definition of homeostasis is:
A.	a condition of remaining the same
B.	all for one and one for all
C.	the maintenance of a relatively stable internal environment
D.	likes repel, opposites attract
E.	what goes up must come down

_____36.	This non-graded membrane potential opens electrically operated sodium channel gates creating the start of a nerve impulse. It is the:
A.	excitatory postsynaptic potential
B.	inhibitory postsynaptic potential
C.	end plate potential
D.	threshold potential
E.	executive potential

_____37.	Which of the following maintains cell shape and mediates cell movements:
A.	the cytoskeleton (microtubules and microfilaments)
B.	mitochondria
C.	endoplasmic reticula
D.	the centrosome
E.	lysosomes

_____38.	Given a bad name by the popular press, _____ is a vital molecule used by the body to make hormones and strengthen the lipid bilayer of the plasma membrane. It is:
A.	adenosine triphosphate
B.	cyclic AMP
C.	cholesterol
D.	nicotine
E.	sugar

_____39.	All systems function in an interrelated fashion in order to preserve the environment of your body in a relatively stable state. This defines:
A.	osmosis	D.	anabolism
B.	catabolism	E.	metabolism
C.	homeostasis

_____40.   Molecules in the outer half of the plasma membrane that are part fat and part sugar and function in the cellular reception of signals are the:
   A.   phospholipids
   B.   glycoproteins
   C.   ectoproteins
   D.   glycolipids
   E.   cholesterol

_____41.   Given a bad name by the popular press, _____ is a vital molecule used by the body to make hormones and strengthen the lipid bilayer of the plasma membrane. It is:
   A.   lipoprotein
   B.   benzine
   C.   cholesterol
   D.   alcohol
   E.   sugar

_____42.   This organelle has a "trans" face and a "cis" face. It is called:
   A.   the nucleus
   B.   the smooth endoplasmic reticulum
   C.   the golgi apparatus
   D.   a lysosome
   E.   a mitochondrion

_____43.   Crenation of a red blood cell would occur if the cell were placed in a _____ solution.
   A.   hypotonic salt
   B.   isotonic salt
   C.   hypertonic salt
   D.   distilled water
   E.   two of the preceding

_____44.   Regarding skeletal muscles, it is true that:
   A.   during isotonic contraction, muscle length does not change
   B.   during isometric contraction, muscle tension does not change
   C.   during isotonic contraction, the muscle performs no mechanical work
   D.   when a muscle is resting at optimal length it is capable of being stimulated to perform maximal work
   E.   normal contractions of skeletal muscle (such as in walking) are either isotonic or isometric but never mixed

_____45.　　Excitatory post-synaptic potentials (EPSP's) occur when a neurotransmitter engages a post-synaptic membrane receptor, resulting in the membrane:

　　A.　　becoming more permeable to Cl^- in
　　B.　　becoming more permeable to K^+ in
　　C.　　hyperpolarizing
　　D.　　releasing acetylcholine
　　E.　　becoming more permeable to Na^+ in

_____46.　　Which of the following is the third of five steps in the scientific method?

　　A.　　hypothesis　　　C.　　publication　　　E.　　experiment
　　B.　　observation　　　D.　　conclusion

_____47.　　Which of the following requires a supply of metabolic energy because the solute moves "uphill"?

　　A.　　dialysis
　　B.　　osmosis
　　C.　　facilitative diffusion
　　D.　　net diffusion
　　E.　　primary active transport

_____48.　　The Na^+ - K^+ pump in the plasma membrane:

　　A.　　is an example of symport
　　B.　　pumps Na^+ in and K^+ out of the cell
　　C.　　generates but does not maintain the ER
　　D.　　is an example of antiport
　　E.　　sometimes but not always is energy dependent

_____49.　　Facilitated diffusion:

　　A.　　always requires a transport protein (carrier)
　　B.　　does not require a membrane potential
　　C.　　does not require a membrane
　　D.　　involves "up-hill" transport of solvent
　　E.　　two of the preceding

_____50.　　Which of the following would be most likely to pass through the plasma membrane (from outside to inside the cell). Assume equal concentration gradients, membrane surface area, membrane thickness, and temperature.

　　A.　　positively charged, lipid insoluble, no carrier
　　B.　　high molecular weight, lipid soluble, no carrier
　　C.　　lipid soluble, low molecular weight
　　D.　　lipid insoluble, high molecular weight
　　E.　　positively charged, low molecular weight, carrier

_____51.   Phospholipids often form _____ to "hide" their hydrophobic tails.
   A.   micelles
   B.   bilayers
   C.   spirals
   D.   helices
   E.   two of the preceding

_____52.   A secretory vesicle is produced by the:
   A.   nucleus
   B.   golgi complex
   C.   lysosomes
   D.   microtubules
   E.   microfilaments

_____53.   This molecule, an integral part of the plasma membrane, is oriented so that its nonpolar part projects to the center of the membrane and its polar part projects toward the surface of the membrane. It is:
   A.   an integral protein
   B.   an ectoprotein
   C.   a glycolipid
   D.   a peripheral protein
   E.   a glycoprotein

_____54.   Physiologically, the term homeostasis means:
   A.   everything's the same inside the body
   B.   nothing ever changes inside the body
   C.   the body's internal environment is normally stable
   D.   to prefer one's own sex
   E.   to mix body fluids until they are uniform

_____55.   Physiology is:
   A.   the study of body structure
   B.   the study of body chemistry
   C.   the study of earth, air, fire, and water
   D.   the study of the functions of living systems
   E.   the study of cells

_____56.   Which of the following is most related to the mitochondrion?
   A.   synthesis of lipids (fats)
   B.   transformation of energy
   C.   intracellular digestion
   D.   reception of extracellular signals
   E.   intracellular circulation

_____57.   This molecule, forming the bilayer core of the plasma membrane, is oriented so that its nonpolar part projects to the center of the membrane and its polar part projects toward the surface of the membrane. It is:
A.   an integral protein
B.   an ectoprotein
C.   a phospholipid
D.   a peripheral protein
E.   a glycoprotein

_____58.   The net diffusion of a solute through a selectively permeable membrane defines:
A.   primary active transport
B.   dialysis
C.   osmosis
D.   antiport
E.   symport

_____59.   Homeostasis refers to:
A.   mechanisms that prevent blood loss
B.   chemical reactions that degrade large molecules
C.   the maintenance of a relatively stable internal environment of the body
D.   chemical reactions that are constructive or synthetic
E.   the study of nature

_____60.   The rate at which a non-polar solute enters a cell is directly proportional to:
A.   its molecular weight
B.   its molecular charge
C.   its molecular size
D.   its concentration difference
E.   water solubility

_____61.   This organelle is basically a package of digestive enzymes which are used intracellularly. It is the:
A.   mitochondrion
B.   golgi apparatus
C.   lysosome
D.   endoplasmic reticulum
E.   centrosome

_____62. Which of the following generalized chemical reactions is the best example of catabolism?

A.   X + Y   ⇒   XY
B.   XY + Z ⇒ X + Y + Z
C.   XY + Z ⇒   XZ + Y
D.   XZ + Y ⇒   XY + Z
E.   ZY + X ⇒   XY + Z

_____63. Because of the typical electrical gradient that exists across plasma membranes, a positive ion (cation) would tend:

A.   to leave the cell
B.   to be unable to pass through the plasma membrane
C.   to enter the cell
D.   to be neutralized
E.   to become an anion

_____64. Acetylcholinesterase:

A.   is the neurotransmitter always found at the neuromuscular junction
B.   is a very common neurotransmitter and is found in central and peripheral nervous systems
C.   is an enzyme which cleaves acetylcholine from the receptor and then splits the molecule into acetyl and choline portions
D.   is an enzyme essential in the synthesis of acetylcholine
E.   has the same effect as curare

_____65. The organelle where proteins destined for export by the cell (secretion) are packaged and activated is called the:

A.   lysosome   D.   ribosome
B.   mitochondrion   E.   golgi apparatus
C.   nucleolus

_____66. A nerve fiber has a resting potential of –80mv and a threshold potential of –55mv. Which of the following is most likely to generate an action potential?

A.   +10mv EPSP
B.   -15mv IPSP
C.   20mv IPSP
D.   +15mv EPSP
E.   A + D (sum)

_____67.   Relative to the transmission of a nerve impulse, the membrane potential at which the membrane suddenly becomes highly permeable to Na^+ is termed the:
A.   action potential
B.   resting potential
C.   threshold potential
D.   EPSP
E.   IPSP

_____68.   Which of the statements regarding plasma membranes is false?
A.   The polar parts of the lipid components are located on the inner and outer surfaces of the plasma membrane.
B.   The nonpolar core of the plasma membrane forms a significant barrier to materials crossing the membrane.
C.   A number of proteins in the plasma membrane form pores or channels that allow some materials to pass through.
D.   Chemical residues associated with reception of information are found on the inner surface of the plasma membrane.
E.   Cholesterol is a component of plasma membranes.

_____69.   Which of the following characterizes synthetic or constructive chemical reactions within the cell?
A.   autolysis
B.   catabolism
C.   metabolism
D.   anabolism
E.   excretion

_____70.   The rate at which a nonpolar solute molecule enters a cell is dependent on:
A.   its solubility in lipids
B.   its molecular size
C.   its molecular charge
D.   both A and B
E.   A, B, and C

_____71.   Depolarization of a dendrite results in current flow between the dendrite and the _____ resulting in propagation of a nerve impulse.
A.   adjacent dendrites
B.   terminal boutons
C.   perikaryon
D.   axon hillock
E.   myelin sheath

_____72.   The plasma membrane of a cell:
   A.   behaves as though the inner surface of the membrane is lined with positive charges and the outer surface as though it were covered with negative charges
   B.   tends to cause a higher concentration of Na^+ inside the cell than outside
   C.   tends to cause a higher concentration of K^+ outside the cell than inside
   D.   tends to cause a higher anion (mostly protein) concentration outside the cell than inside
   E.   tends to be impermeable to anions which are mostly protein

_____73.   Which of the following molecules stabilizes the lipid bilayer of the plasma membrane?
   A.   cholesterol
   B.   phosphate
   C.   ectoprotein
   D.   glycoprotein
   E.   glucose

REFER TO THE FOLLOWING DIAGRAM AND ANSWER QUESTIONS 74, 75, 76

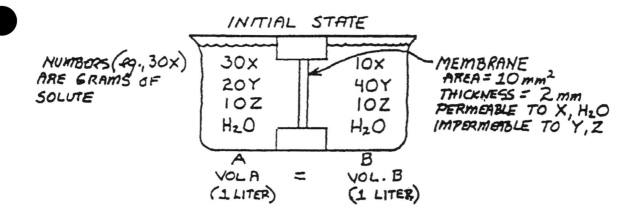

_____74.   The concentration gradient for molecule X is:
   A.   20gms/liter/mm
   B.   20gms/500ml/mm
   C.   10gms/liter/mm
   D.   10gms/500ml/mm
   E.   5gms/liter/mm

_____75. In the initial state:
- A. water is more concentrated in solution A
- B. water is more concentrated in solution B
- C. solutions A and B would be iso-osmotic if the membrane were impermeable to XYZ
- D. water will net diffuse from B to A
- E. none of the preceding statements are true

_____76. At diffusion equilibrium:
- A. compartment A will contain more water (i.e., the level will have risen)
- B. compartment A will contain more Y
- C. compartment B will contain more water
- D. compartment A and B will contain the same amounts of X, Y, Z
- E. the water level in compartments A and B will be the same

_____77. By way of the sodium-potassium pump:
- A. Na^+ and K^+ both leave the cell
- B. Na^+ and K^+ both enter the cell
- C. Na^+ leaves and K^+ enters the cell
- D. K^+ leaves and Na^+ enters the cell
- E. glucose leaves the cell

_____78. Human erythrocytes (red blood cells) placed in a hypertonic salt solution (e.g. 1.2% NaCl) will:
- A. crenate
- B. hemolyze
- C. swell
- D. lose hemoglobin
- E. remain normal in size and shape

_____79. Primary synthesis of a protein hormone secreted by an endocrine cell occurs in the:
- A. lysosome
- B. golgi apparatus
- C. mitochondrion
- D. rough endoplasmic reticulum
- E. centrosome

_____80. Integration of incoming signals is a function of which part of the neuron?
- A. Dendrites
- B. Terminal boutons
- C. Axon
- D. Perikaryon
- E. Nucleus

_____81.　Excitatory post-synaptic potentials occur when a neurotransmitter engages a post-synaptic membrane receptor, resulting in the membrane:

A.　becoming more permeable to Cl^- in
B.　becoming more permeable to K^+ in
C.　becoming more permeable to Na^+ in
D.　becoming less permeable to K^+ in
E.　becoming less permeable to Na^+ in

_____82.　A nerve fiber has a resting potential of -70mv and a threshold potential of -55mv. Which of the following is most likely to generate an action potential?

A.　+10mv EPSP
B.　-15mv IPSP
C.　+15mv EPSP
D.　-20mv IPSP
E.　A + B (sum)

_____83.　The concentration of X in area A is 30 mgm/ml. The concentration of X in area B is 10 mgm/ml. The length of the diffusion pathway that connects areas A and B is 20 centimeters. The concentration gradient for the net diffusion of X from A to B is:

A.　20 mgm/ml/cm
B.　10 mgm/cm
C.　1 ml/cm/
D.　1 mgm/ml/cm
E.　40 mgm/cm

_____84.　The principal function of this organelle is to generate high energy phosphate compounds that provide usable energy for cell work. It is the:

A.　smooth ER
B.　golgi complex
C.　mitochondrion
D.　lysosome
E.　nucleolus

_____85.　Plasma membrane transport molecules (carriers):

A.　are usually lipids
B.　generally transport many types of solutes (i.e., the carriers are not very specific)
C.　are proteins that remain confined to the plasma membrane
D.　transport water as well as solute
E.　are not involved in facilitated diffusion

_____86.   What is the direction of the driving forces for the movement of sodium ions when a nerve cell is at rest?

A.   inward electrical gradient
B.   outward electrical gradient
C.   inward chemical gradient
D.   both A and C
E.   both B and C

_____87.   A cigar-shaped membranous organelle that contains enzymes used to generate ATP is:

A.   the smooth E.R
B.   a microtubule
C.   a mitochondrion
D.   the nucleus
E.   a lysosome

_____88.   Chemical changes resulting in the building up of living material are described by the term:

A.   catabolism
B.   osmosis
C.   anabolism
D.   autolysis
E.   metabolism

_____89.   The concentration of A in area X is 40 mg/ml. The concentration of A in area Y is 20 mg/ml. The length of the diffusion pathway that connects area X to area Y is 20 millimeters. The concentration gradient of the net diffusion of A from X to Y is:

A.   20 mg/mm   D.   1 mg/ml/mm
B.   10 mg/ml/mm   E.   1 ml/mm
C.   zero

_____90.   The individual known for pneumatology was:

A.   Galen
B.   Harvey
C.   Malpighi
D.   Vesalius
E.   Priestly

_____91.   The type of molecule that would have the most difficulty diffusing through the lipid bilayer of the plasma membrane would be a molecule that has the following combination of characteristics:

A.   polar, low lipid solubility, large molecule
B.   non-polar, high lipid solubility, small molecule
C.   polar, lipid insoluble, small molecule
D.   non-polar, low lipid solubility, large molecule
E.   small, positively charged, low lipid solubility

_____92.   The type of molecule most likely to enter a cell by diffusing through a protein channel in the plasma membrane would be:
A.   large and positively charged
B.   small and negatively charged
C.   small, and positively charged
D.   large and polar
E.   soluble in lipid

_____93.   Which statement is true?
A.   EPSP's always generate action potentials.
B.   Acetylcholine is always used as a neurotransmitter at the neuromuscular junction.
C.   IPSP's never generate action potentials.
D.   Norepinephrine is always used as a neurotransmitter at the neuromuscular junction.
E.   Two of the preceding are true.

_____94.   Symport refers to the active transport of a solute across the plasma membrane:
A.   into the cell
B.   in the same direction as Na^+
C.   in the opposite direction as Na^+
D.   coupled to Na^+
E.   answers B and D

_____95.   One example of a non-membranous organelle is the:
A.   lysosome
B.   golgi apparatus
C.   mitochondrion
D.   nucleus
E.   centrosome

_____96.   Following are five steps in the scientific method. Which step is normally the second step?
A.   hypothesis
B.   observation
C.   experiment
D.   publication
E.   conclusion

_____97.   A nerve fiber has a resting potential of –70mv and a threshold potential of –55mv. Which of the following is most likely to generate an action potential?
A.   +10mv EPSP   D.   +20mv EPSP
B.   -15mv IPSP   E.   A or C
C.   +5mv EPSP

_____98.   The process by which secretory vesicles fuse with the plasma membrane and expel their contents outside the cell is called:

    A.   active transport
    B.   phagocytosis
    C.   exocytosis
    D.   pinocytosis
    E.   endocytosis

_____99.   Inhibitory post-synaptic potentials occur when a neurotransmitter engages a post-synaptic membrane receptor, resulting in the membrane:

    A.   becoming more permeable to K^+out
    B.   becoming more permeable to Na^+in
    C.   becoming less permeable to K^+out
    D.   becoming less permeable to Na^+out
    E.   allowing Cl^- to leak out

_____100.   Examine the following diagram and answer questions 101, 102, l03 and 104.

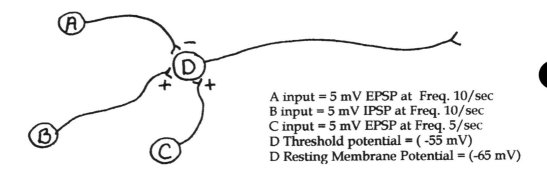

A input = 5 mV EPSP at Freq. 10/sec
B input = 5 mV IPSP at Freq. 10/sec
C input = 5 mV EPSP at Freq. 5/sec
D Threshold potential = (-55 mV)
D Resting Membrane Potential = (-65 mV)

_____101.   Neuron D will fire (generate and conduct a nerve impulse) if:

    A.   inputs A and B occur simultaneously at frequency shown
    B.   inputs B and C occur simultaneously at frequency shown
    C.   inputs A, B and C occur simultaneously at frequency shown
    D.   inputs A and C occur simultaneously at frequency shown
    E.   two of the preceding

_____102.   When Neuron D integrates simultaneous inputs from B and C, the process is called:

    A.   cancellation
    B.   temporal summation
    C.   spatial summation
    D.   depolarization
    E.   refractory

_____103.   If Neuron A were to fire several times in rapid succession, Neuron D
              would fire. The process whereby Neuron D integrates this input is
              called:
              A.   spatial summation
              B.   temporal summation
              C.   repolarization
              D.   depolarization
              E.   frequency coding

_____104.   Which of the cells is an effector?
              A.   neuron A
              B.   neuron B
              C.   neuron C
              D.   neuron D
              E.   none of the above

MATCHING. CHOOSE THE BEST ANSWER. ANSWERS MAY BE USED MORE THAN ONCE OR NOT AT ALL.

              A.   organelle where most ATP is synthesized
              B.   packages proteins prior to their secretion
              C.   contains enzymes used in intracellular digestion
              D.   organelle where most intracellular proteins are synthesized
              E.   contains the genetic blueprints that govern protein synthesis

_____105.   Golgi apparatus

_____106.   nucleus

_____107.   mitochondrion

_____108.   rough endoplasmic reticulum

_____109.   lysosome

A. graded; nonpropagated; increased chloride ion conductance into postsynaptic cell
B. nongraded; propagated; all or none law
C. nonpropagated; graded; increased sodium ion conductance into postsynaptic cell which is not skeletal muscle
D. motor end-plate; increased sodium ion conductance into postsynamptic cell
E. stable; maintained by Na^+ - K^+ pump

_____110. end-plate potential

_____111. inhibitory postsynaptic potential

_____112. excitatory postsynaptic potential

_____113. action potential

_____114. resting membrane potential

A. mitochondrion
B. golgi Apparatus
C. lysosome
D. smooth E.R
E. rough E.R

_____115. combines with phagosome to form digestive vacuole

_____116. primary site of adenosine triphosphate synthesis

_____117. associated with ribosomes and protein synthesis

_____118. synthesizes lipids (fats)

_____119. packages cell secretions and forms secretory vesicles

A. energy dependent carrier-mediated transport
B. net diffusion
C. facilitative diffusion
D. osmosis
E. dialysis

_____120. requires a membrane; is directly coupled to an energy-yielding reaction; moves solute against gradient

_____121. solute movement from higher to lower solute concentration without membrane

_____122. solvent movement across a membrane from higher to lower solvent concentration

_____123. solute movement by carrier across membrane from higher to lower solute concentration

_____124. net diffusion of solute across a selectively permeable membrane

A. excitatory post-synaptic potential
B. end-plate potential
C. action potential
D. inhibitory post-synaptic potential
E. threshold potential

_____125. the most negative (inside cell with respect to outside cell) of all the membrane potentials listed above

_____126. a non-graded potential that is propagated in an all or none fashion

_____127. a membrane potential that opens electrically operated sodium gates in the nerve cell plasma membrane

_____128. a graded potential (not all or none) measured at the neuromuscular junction

_____129. a variable amplitude potential that increases the chance of the axon hillock (initial segment) generating a nerve impulse

A. Graded. Excitatory. Not conducted in all or none manner. Occurs
   At neuromuscular junction.
B. Not graded. Not conducted in all or none manner. Opens
   Electrically operated sodium gates.
C. Graded. Excitatory. Not conducted in all or none manner. Opens
   Chemically operated sodium gates in post-synaptic neuron
   Membrane.
D. Graded. Inhibitory. Not conducted in all or none manner.
   Increases chloride ion movement into post-synaptic neuron.
E. Non-graded. All or none propagation.

_____130. E.P.S.P

_____131. A.P.

_____132. I.P.S.P.

_____133. E.P.

_____134. T_r

A. Mitochondrion
B. Lysosome
C. Rough Endoplasmic Reticulum
D. Nucleus
E. Golgi Olgi Apparatus

_____135. This organelle is basically a package of digestive enzymes which are used
intracellularly.

_____136. The principle function of this organelle is energy transformation and the
manufacture of high-energy phosphates.

_____137. A protein hormone made by the cell is packaged in final form for
secretion by this organelle.

_____138. Associated with ribosomes and protein synthesis. Most proteins made by
the cell are put together in this organelle.

_____139. Contains the genetic blueprints for protein synthesis.

ENCIRCLE THE LARGER ITEM IN EACH OF THE FOLLOWING PAIRS OF ITEMS. IF NEITHER ITEM IS LARGER, ENCIRCLE C.

A B C 140. The strength of skeletal muscle contraction when the number of motor units activated is:

A.   increased
B.   decreased

A B C 141. The strength of skeletal muscle fiber contraction when the fiber is stimulated with a single:

A.   maximal stimulus (above threshold)
B.   threshold stimulus

A B C 142. Mechanical work performed by skeletal muscles during:

A.   isometric contraction
B.   isotonic contraction

A B C 143. Diffusion of sodium into the neuron:

A.   during depolarization
B.   during repolarization

A B C 144. Calcium ion release from the sarcoplasmic reticulum:

A.   during sarcomere contraction
B.   during sarcomere relaxation

PHYSIOLOGY OUTLINES 7-12

MULTIPLE CHOICE. SELECT THE BEST ANSWER.

_____145.   Deviation of the tongue to the left suggests damage to the:
    A.   right C.N. XII
    B.   left C.N. V
    C.   left C.N. IX
    D.   left C.N. XII
    E.   left C. N. VII

_____146.   Damage to the lateral spinothalamic pathway in the left brainstem could result in:
    A.   loss of cutaneous pain on the left side of the trunk
    B.   loss of light touch from the left arm skin
    C.   loss of cutaneous temperature sensation from the right leg
    D.   dilated left pupil
    E.   spastic paralysis of the arm

_____147.   The motor division of cranial nerve V on the right side of the brainstem has become nonfunctional. As a result, the subject has lost entirely or partially the ability to:
    A.   raise the right eyebrow
    B.   frown
    C.   clench teeth
    D.   sense pain from the right cheek skin
    E.   shed tears

_____148.   Extension of the leg involves:
    A.   inhibition of extensor gamma motorneurons
    B.   stimulation of flexor gamma motorneurons
    C.   stimulation of extensor alpha motorneurons
    D.   stimulation of flexor alpha motorneurons
    E.   two of the preceding

_____149.   The primary purpose or function of the gamma motorneuron is to:
    A.   prevent flexors and extensors from contracting at the same time
    B.   facilitate the withdrawal reflex
    C.   maintain posture of the body
    D.   adjust sensitivity of the neuromuscular spindle
    E.   inhibit the alpha motorneuron

_____150. Contraction of the ciliary muscle causes:
   A.   the eyeball to rotate laterally
   B.   the pupil to become smaller
   C.   the lens to become thinner (flatten)
   D.   the lens to become thicker (bulge)
   E.   the near-point to recede

_____151. According to the motor homunculus (disproportionate human figure), the largest area of motor control by the cerebral cortex is dedicated to controlling muscles of the _____.
   A.   toes
   B.   legs
   C.   arms
   D.   thumb and fingers
   E.   anterior and posterior trunk

_____152. The part of the body with the greatest sensitivity to somatic sensory stimuli, and therefore the part with the greatest representation in the post central gyrus is (are) the:
   A.   arm
   B.   hand
   C.   fingers and thumb
   D.   toes
   E.   hips

_____153. Blocking nicotinic autonomic receptors would result in:
   A.   decreasing sympathetic but not parasympathetic activity
   B.   decreasing parasympathetic but not sympathetic activity
   C.   increasing sympathetic and decreasing parasympathetic activity
   D.   increasing parasympathetic and decreasing sympathetic activity
   E.   decreasing both parasympathetic and sympathetic activity

_____154. Damage to the dorsal columns, such as may occur in tabes dorsalis (syphilis) would result in partial or complete:
   A.   loss of cutaneous two point discrimination
   B.   loss of skeletal muscle coordination
   C.   loss of pain sensation
   D.   loss of temperature sensation
   E.   answers (a) and (b)

_____155. Interpreting sound intensity (loudness) involves:
   A.   the parietal cerebrum
   B.   the hypothalamus
   C.   the organ of Corti
   D.   the utricle
   E.   two of the preceding

_____156. Primary control of heart rate and blood pressure is a function of the:
   A. thalamus
   B. mesecephalon
   C. tenth cranial nerve
   D. medulla oblongata
   E. two of the preceding

_____157. A blindfolded patient can feel or sense the weight of an object as well as its texture (rough, smooth, etc.) when placed in his/her right hand but cannot identify its shape (cube, sphere, disc, etc.) based on how it feels. A probable area of brain damage is the:
   A. left frontal association cortex
   B. left parietal association cortex
   C. right temporal association cortex
   D. right occipital association cortex
   E. left hypothalamus

_____158. Autonomic alpha-one adrenergic receptors are:
   A. blocked by nicotine
   B. sympathetic receptors that mediate contraction of vascular smooth muscle
   C. stimulated by alpha-two receptors
   D. excitatory everywhere except the heart
   E. two of the preceding

_____159. Which of the following is most likely to be associated with a lesion (area of damage) of the frontal cortex?
   A. partial anesthesia
   B. flaccid paralysis
   C. altered mood
   D. spastic paralysis
   E. partial deafness

_____160. Which reflex is monosynaptic and ipsilateral?
   A. pupillary dilator
   B. babinski
   C. knee-jerk
   D. withdrawal
   E. two of the preceding

_____161. A blindfolded person can sense the weight of an object as well as its texture when placed in his/her left hand but cannot identify its shape. A probable area of brain damage is the:
   A. left frontal association cortex   E. left hypothalamus
   B. right parietal association cortex
   C. left temporal association cortex
   D. right occipital association cortex

_____162. Damage to the lateral white matter of the spinal cord on the right side of the body between the shoulder blades could result in:
- A. loss of spinal reflexes involving the left arm
- B. loss of spinal reflexes involving the right leg
- C. loss of sensation from the skin of the right leg
- D. paralysis of the right side of the face
- E. sensory but no motor loss involving the left leg

_____163. The modality (type) of a somatic or special sensory stimulus is determined by the brain using a _____ code.
- A. frequency
- B. population
- C. labeled-line
- D. bar
- E. Morse

_____164. Transection (completely cutting through) of the posterior roots of the spinal nerves that supply the posterior thigh would result in:
- A. inability to elicit the knee-jerk reflex
- B. spastic paralysis of posterior thigh muscles
- C. flaccid paralysis of posterior thigh muscles
- D. anesthesia of posterior thigh
- E. two of the preceding

_____165. Upon stepping on a tack, a person lifts the injured foot and extends the opposite leg to maintain balance, the extensor reflex:
- A. is multisynaptic
- B. is ipsilateral
- C. requires the brain for activation
- D. is contralateral
- E. two of the preceding

_____166. Blindfolded, a person is unable to identify the shape of a wooden cube placed in the left hand although the person can sense *when* the object was placed in the hand. The most probable location of a lesion is the:
- A. right temporal association cortex
- B. Broca's area left cortex
- C. right prefrontal cortex
- D. right parietal association cortex
- E. left primary somatic sensory cortex

_____167. Which of the following nerves contains preganglionic sympathetic fibers?
- A. tenth thoracic spinal nerve
- B. trigeminal nerve
- C. eighth cervical spinal nerve
- D. facial nerve
- E. vagus nerve

413

_____168. Autonomic alpha-one adrenergic receptors are:
- A. blocked by norepinephrine
- B. sympathetic receptors that mediate contraction of vascular smooth muscle
- C. stimulated by alpha-two receptors
- D. excitatory everywhere except the heart
- E. extinct

_____169. Which of the following tracts transmits pain information from the skin on the left thigh?
- A. left anterior spinothalamic
- B. right anterior spinothalamic
- C. left lateral spinothalamic
- D. right lateral spinothalamic
- E. left posterior column

_____170. Which of the following is (are) not dominated by the right cerebral hemisphere?
- A. artistic creativity
- B. intuitive thought
- C. subtle meanings of language
- D. logical thought as in science
- E. two or more of the preceding

_____171. If the right eighth thoracic spinal nerve (upper back) were cut, there would be:
- A. a loss of pain sensation from the toes of the left foot
- B. spastic paralysis of right leg muscles
- C. flaccid paralysis of right leg muscles
- D. loss of all reflex activity in body region supplied by RT_8 spinal nerve
- E. two of the preceding

_____172. Damage to the gamma motor neuron system would most likely result in:
- A. spastic paralysis
- B. inability to coordinate muscle activities
- C. flaccid paralysis
- D. loss of myotatic reflexes
- E. two of the preceding

_____173. Which of the following would you expect to observe in a patient with a tumor in the cerebellum?
- A. loss of vision
- B. loss of hearing
- C. inability to perceive pain
- D. inability to initiate voluntary movements
- E. inability to execute smooth, steady movements

_____174. Examine the diagram below. The location of the lesion would result in:

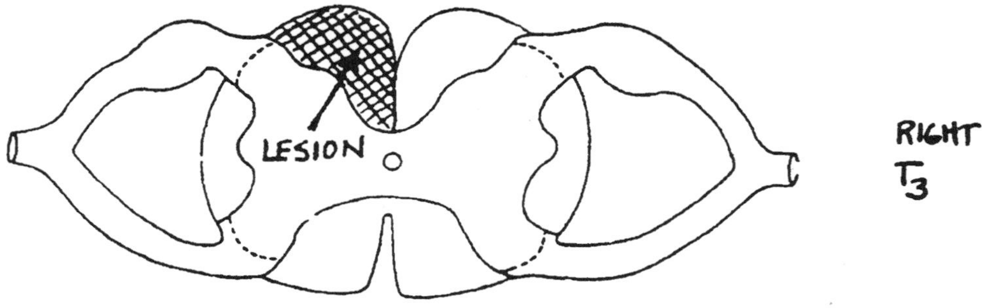

RIGHT
T$_3$

A. complete anesthesia of the right leg
B. spastic paralysis of 85% of the motor units in the left arm
C. flaccid paralysis of 15% of the motor units in the left arm
D. loss of vibratory sensation from the left leg
E. none of the preceding

_____175. In the autonomic nervous system, beta receptors:
A. are cholinergic
B. can be blocked by curare
C. when engaged with transmitter, increase heart rate
D. are found only in skeletal muscle
E. are muscarinic

_____176. Damage to the left anterolateral pathway at spinal segment T$_{10}$ could result in:
A. loss of conscious muscle position in the right leg
B. loss of fine skilled movement in the left leg
C. loss of cutaneous pain sensation from the right leg
D. loss of cutaneous pressure sensation from the left leg
E. a sensory loss other than one of the above

_____177. Which type of receptor is found on target cells of preganglionic parasympathetic nerve fibers?
A. beta
B. alpha
C. muscarinic
D. curariform
E. nicotinic

_____178. That portion of the brainstem concerned with visual reflexes such as the pupillary reflex is the:
A. pons
B. mesencephalon
C. medulla oblongata
D. thalamus
E. hypothalamus

_____179.   Damage to cranial nerve V would most likely interfere with:
　　　　　　A.　vision
　　　　　　B.　hearing
　　　　　　C.　swallowing
　　　　　　D.　chewing
　　　　　　E.　balance

_____180.   Cutaneous two-point discrimination requires:
　　　　　　A.　lateral spinothalamic tracts
　　　　　　B.　anterior corticospinal tracts
　　　　　　C.　premotor cortex
　　　　　　D.　posterior columns
　　　　　　E.　anterior spinocerebellar tracts

_____181.   Damage to the following nerve may result in paralysis of facial muscles:
　　　　　　A.　C.N. III
　　　　　　B.　C.N. V
　　　　　　C.　C.N. VII
　　　　　　D.　C.N. X
　　　　　　E.　C.N. VI

_____182.   An application of pressure to the eyeball results in a visual sensation even though the usual stimulus modality for photoreceptors is light. The experiment demonstrates the law of:
　　　　　　A.　neutrality　　　　　　　　　　E.　Newton
　　　　　　B.　adequate stimulus
　　　　　　C.　all or none
　　　　　　D.　specific nerve energies

_____183.   The right optic tract has been compressed by a tumor, blocking transmission in the optic fibers. The loss of vision:
　　　　　　A.　would exclusively involve the right eye
　　　　　　B.　would involve the visual fields of both eyes
　　　　　　C.　would involve the visual cortex of both right and left cerebral hemispheres
　　　　　　D.　would result in complete blindness
　　　　　　E.　would be called emmetropia

_____184.   One of the following statements applies to the sympathetic division of the ANS. Which one?
　　　　　　A.　The ganglia are usually within walls of viscera.
　　　　　　B.　The receptor for postganglionic transmitter is muscarinic.
　　　　　　C.　None of the receptors for transmitters are nicotinic.
　　　　　　D.　Atropine blocks the receptor for postganglionic neurotransmitter.
　　　　　　E.　Activation of the beta receptor usually causes smooth muscle to relax.

_____185. Damage to alpha motorneurons may result in:
A. spastic paralysis of skeletal muscle
B. paralysis of smooth muscle
C. loss of awareness concerning muscle strength
D. flaccid paralysis of skeletal muscle
E. two of the preceding

_____186. Damage to the lateral white matter of the spinal cord on the right side of the body between the shoulder blades could result in:
A. loss of spinal reflexes involving the left arm
B. loss of spinal reflexes involving the right leg
C. loss of sensation from the skin of the right leg
D. paralysis of the right side of the face
E. sensory but no motor loss involving the left leg

_____187. Which nerve controls heart rate, gastric secretion, and pancreatic secretion?
A. C.N. X
B. C.N. IX
C. C.N. VII
D. C.N. XII
E. C.N. XI

_____188. Damage to the dorsal columns, such as may occur in tabes dorsalis (syphilis) would result in partial or complete:
A. loss of vibratory sensation
B. loss of skeletal muscle coordination
C. loss of pain sensation
D. loss of temperature sensation
E. answers A and B

_____189. Which of the following nerves contain preganglionic sympathetic fibers:
A. tenth thoracic spinal nerve
B. trigeminal nerve
C. eighth cervical nerve
D. facial nerve
E. optic nerve

_____190. Rotation of the eyeball in the orbit requires cranial nerves _____.
A. 5-6-7
B. 2-3-5
C. 3-4-6
D. 2-4-6
E. 3-4-5

_____191. A loss of taste and salivation could result from damage to the _____ cranial nerve.
A. fifth
B. twelfth
C. second
D. seventh
E. none of the preceding

_____192. The ipsilateral withdrawal reflex:
A. is a monosynaptic reflex arc
B. requires association or interneurons
C. simultaneously activates ipsilateral extensors and flexors
D. can easily be voluntarily inhibited once initiated
E. two of the preceding

_____193. A simple spinal reflex such as a withdrawal reflex:
A. is usually multisynaptic
B. is usually ipsilateral
C. usually requires input from the brain
D. always involves anterior and posterior spinal nerve roots
E. all of the above except C

_____194. The most likely result of irregular curvature of the cornea is:
A. myopia
B. hypermetropia
C. emmetropia
D. astigmatism
E. nyctalopia

_____195. Considering the size and number of component motor units and the need for precise control of skeletal muscles, which of the following has the greatest area of representation in the somatic motor cortex?
A. muscles that move the toes
B. muscles that move the legs
C. muscles that move the thumb and fingers
D. muscles that move the arms
E. muscles that move the trunk

_____196. This part of the brain contains the body's thermostat:
A. medulla oblongata
B. pons
C. hypothalamus
D. cerebral cortex
E. cerebellum

_____197. The constant exposure to loud noise near 20,000 Hz would result in loss of hair cells on the basilar membrane:
   A.   near the base of the cochlea
   B.   near the apex of the cochlea
   C.   at a location midway between the base and the apex
   D.   at the helicotrema
   E.   two of the preceding

_____198. All somatic sensory information from the skin of the trunk and extremities is relayed to the somatic sensory cortex by the:
   A.   thalamus
   B.   hypothalamus
   C.   frontal lobe
   D.   temporal lobe
   E.   cerebellum

_____199. Which of the following is not part of the fight or flight response?
   A.   increase in the force of heart contraction
   B.   decrease in blood flow to the skeletal muscles
   C.   increase in pulse rate
   D.   conversion of stored food into glucose
   E.   more oxygen is made available by increased breathing

_____200. Which of the following (is) are dominated by the right cerebral hemisphere?
   A.   artistic creativity
   B.   solving math problems
   C.   motor control of speech
   D.   logical thought
   E.   two of the preceding

_____201. Broca's area of the brain:
   A.   controls hearing
   B.   is usually in the right cerebrum
   C.   controls speech
   D.   involves taste
   E.   two of the preceding

_____202. If your spinal cord was completely severed between C_3 and C_4 vertebrae you would lose:
   A.   pain sensation from the face
   B.   myotatic reflexes of the lower extremities
   C.   myotatic reflexes of the upper extremities
   D.   upper motorneuron control of respiratory muscles
   E.   reflex emptying of the urinary bladder

419

_____203. A patient has suffered a stab wound to the neck resulting in damage to the spinal cord as diagrammed below:

The patient would be unable to:
A. sense cutaneous pain from the right leg
B. voluntarily contract muscles of the right leg
C. sense cutaneous temperature changes from the left leg
D. demonstrate myotatic reflexes of the right leg
E. do two of the above

_____204. This division of the brain controls visual reflexes such as accommodation of the lens. It is the:
A. hypothalamus
B. medulla oblongata
C. pons
D. thalamus
E. mesencephalon

_____205. Localization and differentiation of thermal vs. mechanical (pressure) cutaneous stimuli involves the _____ cortex of the cerebrum.
A. frontal
B. parietal
C. temporal
D. occipital
E. occipito-temporal

_____206. Stimulation of extensor gamma motorneurons:
A. reduces stretch sensitivity of the opposing flexor
B. increases stretch sensitivity of the extensor
C. causes isometric contraction of the extensor
D. inhibits the contralateral flexor
E. none of the above

_____207. The hypothalamus:
A. controls the internal environment of the body
B. has neuroendocrine fibers
C. controls basic biological rhythms
D. provides motivation for drives
E. all of the above

_____208.   According to the principle of reciprocal innervation:
A.   extensor alpha motorneurons are stimulated when the extensor gamma motorneurons are inhibited
B.   when flexors at a joint are stimulated, extensors at the same joint are stimulated
C.   when flexor alpha motorneurons are stimulated, the alpha motorneurons of the opposing extensor are inhibited
D.   stimulation of ipsilateral flexors is accompanied by contralateral stimulation of extensors
E.   all muscles are supplied by alpha and gamma neurons

_____209.   A patient has suffered a stab wound to the middle of the back (T_7) resulting in damage to the spinal cord as diagrammed below:

RIGHT

The patient would be unable to:
A.   sense cutaneous pain from the left leg
B.   voluntarily contract muscles of the right leg
C.   sense cutaneous temperature changes from the right leg
D.   demonstrate myotatic reflexes of the right leg
E.   do two of the above

_____210.   Before sensory information reaches the cerebral cortex, it is processed and integrated by the:
A.   cerebellum
B.   thalamus
C.   hypothalamus
D.   brainstem
E.   alpha motor neurons

_____211.   Which of the following is (are) dominated by the left cerebral hemisphere?
A.   solving math problems
B.   artistic creativity
C.   intuitive thought
D.   understanding jokes
E.   two of the above

_____212.   When the sympathetic system is activated one sees:
A.   increased blood flow to the skin
B.   increased secretion of gastric enzymes
C.   increased urine production
D.   increased airway diameter
E.   increased production of thin watery saliva

_____213. The following statement is true (if more than one is true, choose E):
A. efferent fibers in the PNS are sensory
B. afferent fibers in the PNS are motor
C. all cranial nerves are classified as mixed
D. all spinal nerves are classified as mixed
E. more than one of the above statements is true

_____214. Daily routine maintenance of visceral activities is the domain of the:
A. primary somatic motor cortex
B. cerebellum
C. premotor cortex in the frontal lobe of the cerebrum
D. sympathetic division of the ANS
E. parasympathetic division of the ANS

_____215. As regards sensory systems, one method of informing the brain about the *intensity* of a stimulus involves a frequency code. Another method involves:
A. a labeled-line code
B. a population code
C. a Morse code
D. the law of Specific Nerve Energies
E. the law of Adequate Stimulus

_____216. This structure issues commands (excitatory and inhibitory) to the gamma motor neurons:
A. primary somatic motor cortex
B. thalamus
C. cerebellum
D. temporal lobe of the cerebrum
E. occipital lobe of the cerebrum

_____217. Which type of receptor is found on target cells of postganglionic parasympathetic nerve fibers?
A. beta receptor
B. alpha receptor
C. muscarinic receptor
D. nicotinic receptor
E. curariform receptor

_____218. Which of the following is (are) true concerning retinal function?
A. Cone vision is best at low levels of light intensity.
B. Color vision is mediated by rods.
C. Rod vision is of low acuity.
D. Visual acuity is highest in the peripheral retina.
E. blind spot = fovea centralis

_____219.   The primary somatic sensory cortex:
  A.   receives information from the organ of Corti
  B.   is located in the occipital lobe of the cerebrum
  C.   perceives pain
  D.   is located in the frontal lobe of the cerebrum
  E.   stores information regarding somatic sensations such as what it feels like to touch a hot stove

_____220.   A patient comes in who has lost control of many muscles on the right side of his body. Strangely, he is able to sense pain, temperature, touch, pressure and vibrations from a tuning fork all over his body on both sides. This patient has most likely suffered:
  A.   an injury in the posterior white columns of the spinal cord
  B.   a cerebral hemorrhage in the left primary motor cortex
  C.   a cerebral hemorrhage in the right primary motor cortex
  D.   injury to the temporal lobe of the cerebrum
  E.   a complete transection of the spinal cord in the cervical (neck) region

_____221.   The structure of the cerebrum that functionally and anatomically connects the two hemispheres is called the:
  A.   thalamus
  B.   corpus callosum
  C.   hypothalamus
  D.   cerebellum
  E.   superior colliculi

_____222.   Cutaneous pain from the right thigh travels to the somatic sensory cortex via the:
  A.   left anterolateral pathway of the spinal cord
  B.   right posterior columns
  C.   left spinocerebellar tract
  D.   anterior corticospinal pathway
  E.   left lateral corticospinal tract

_____223.   Considering the sympathetic division of the autonomic nervous system, which statement is true:
  A.   The preganglionic transmitter is norepinephrine.
  B.   The postganglionic transmitter is acetylcholine except in skeletal muscle.
  C.   The preganglionic neurons are located in the lumbar and sacral spinal cord segments.
  D.   Most postganglionic neurons are very short and located within the walls of viscera.
  E.   None of the preceding statements are true.

224. The primary function of the stapes is to:
   A. relay vibrations of the malleus to the incus
   B. relay vibrations of the tympanic membrane to the malleus
   C. relay vibrations of the incus to the vestibule
   D. dampen excessive vibration of the tympanic membrane
   E. none of the preceding

225. The organ of Corti:
   A. is responsible for equilibrium
   B. is located within the scala tympani
   C. is responsible for hearing
   D. is located within the scala vestibuli
   E. two of the preceding

226. The perception of pain from the skin of the fingers requires normal functioning of the:
   A. frontal cerebral cortex
   B. anterior spinothalamic tracts
   C. parietal cerebral cortex
   D. lateral spinothalamic tracts
   E. two of the preceding

227. The organ of Corti is:
   A. located in the scala media
   B. responsible for hearing
   C. responsible for equilibrium
   D. located in the scala tympani
   E. two of the preceding

228. The trichromatic theory of color vision is based on:
   A. rods, not cones
   B. red rods, blue cones
   C. red, blue, and green cones
   D. red, white and blue cones
   E. Dr. Shinobu Ishihara

229. Interpretation of the magnitude or intensity of a pressure stimulus applied to the skin involves:
   A. the amplitude of the sensory nerve fiber action potentials
   B. the number of sensory nerve fibers activated simultaneously
   C. lateral spinothalamic tracts
   D. parieto-occipital association cortex
   E. two of the preceding

_____230. You reach to pick up an object on a table. The command to pick up was issued by the _____ and the part of the brain ensuring coordination of appropriate muscles and overall smoothness of the movement is the _____.
A.   medulla, pons
B.   cerebellum, cerebrum
C.   thalamus, hypothalamus
D.   cerebrum, mesencephalon
E.   cerebrum, cerebellum

_____231. Flexion of the forearm is accompanied by:
A.   stimulation (+) of extensor alpha motorneurons
B.   stimulation (+) of extensor gamma motorneurons
C.   inhibition (-) of flexor alpha motorneurons
D.   inhibition (-) of flexor gamma motorneurons
E.   inhibition (-) of extensor gamma motorneurons

_____232. The right corner of the mouth and the right upper eyelid droop. there is noticeable lack of muscle tone in the right cheek. This person suffers from damage to:
A.   left C.N. V   D.   left C.N. IX
B.   right C.N. VII   E.   right C.N. VI
C.   right C.N. III

_____233. Which of the following reflects increased sympathetic activity?
A.   vasodilation in skeletal muscle
B.   decreased airway diameter
C.   decreased heart rate
D.   pupillary constriction
E.   two of the preceding

_____234. Upon stepping on a tack, a person lifts the injured foot and extends the opposite leg to maintain balance. The extensor reflex:
A.   is monosynaptic
B.   is ipsilateral
C.   requires the brain for activation
D.   is contralateral
E.   two of the preceding

_____235. Damage to the lateral white matter of the spinal cord on the right side of the body between the shoulder blades could result in:
A.   loss of spinal reflexes involving the left arm
B.   loss of spinal reflexes involving the right leg
C.   loss of sensation from the skin of the left leg
D.   paralysis of the right side of the face
E.   sensory but not motor loss involving the right leg

_____236. Which of the following is associated with damage to the cerebellum?
  A.   spastic paralysis
  B.   flaccid paralysis
  C.   past-pointing
  D.   inability to sense body position
  E.   loss of myotatic reflexes

_____237. The function of the ciliary muscle of the eye is:
  A.   convergence
  B.   to adjust the amount of light entering the eye
  C.   to distinguish between light and dark objects
  D.   to distinguish different colors
  E.   none of the preceding

_____238. Alpha receptors:
  A.   are cholinergic
  B.   are muscarinic
  C.   are nicotinic
  D.   are adrenergic
  E.   can be blocked by curare

_____239. Which of the following functions is associated with the parietal lobe of the cerebrum?
  A.   initiation of skeletal muscle contraction
  B.   recognition of sound frequency and amplitude
  C.   discrimination between thermal and mechanical stimuli
  D.   two of the preceding
  E.   none of the preceding

_____240. As in walking, when the brain initiates contraction of a flexor muscle via the alpha motor neuron, it also:
  A.   increases gamma activity in the opposing extensor
  B.   decreases gamma activity in the opposing extensor
  C.   decreases activity in the flexor
  D.   increases alpha activity in the extensor
  E.   does none of the above

_____241. If the sensory portion of this nerve were cut, facial anesthesia would result:
  A.   C.N. III
  B.   C.N. V
  C.   C.N. II
  D.   C.N. IX
  E.   C.N. VI

_____242. Sensory receptors are most sensitive to one modality of stimulus. This is known as the law of:
A. Murphy
B. specific nerve energies
C. adequate stimulus
D. threshold stimulus
E. ESP

_____243. The ability to describe the shape of an object held in the hand by a blindfolded subject requires normal functioning of the subject's:
A. occipital cerebral cortex
B. prefrontal cerebral cortex
C. tectum of the mesencephalon
D. parietal cerebral cortex
E. hypothalamus

_____244. Which of the following functions is associated with the frontal lobe of the cerebrum?
A. initiation of skeletal muscle contraction
B. recognition of sound frequency and amplitude
C. discrimination between thermal and mechanical stimuli
D. discrimination between visual and auditory stimuli
E. two of the preceding

_____245. Which of the following motor activities best exemplifies the case where there is a *simultaneous* increase in gamma motorneuron activity to both agonist and antagonist muscles at a given joint?
A. walking
B. running
C. swimming
D. standing at attention
E. lifting barbells

_____246. With respect to simple spinal reflexes:
A. all reflex arcs are multisynaptic
B. all reflex arcs involve a sensory, internuncial, and motor neuron
C. all neurotransmitters used in synaptic transmission are excitatory
D. all sensory information is relayed to higher neural centers
E. none of the preceding

_____247. Which of the following describes autonomic tone?
A. The sympathetics excite; the parasympathetics inhibit.
B. Most viscera have sympathetic and parasympathetic innervation.
C. The two divisions of the ANS usually oppose each other and are always active, although their levels of activity vary.
D. Alpha and beta receptors have opposite effects.
E. Two of the preceding.

_____248. Examine the diagram below. The location of the lesion would result in:

RIGHT T₃

A. complete anesthesia of the right leg
B. spastic paralysis of 85% of the motor units in the left arm
C. flaccid paralysis of 15% of the motor units in the left arm
D. loss of vibratory sensation from the right leg
E. none of the preceding

_____249. Primary control of heart rate and blood pressure is a function of the:
A. thalamus
B. mesencephalon
C. fifth cranial nerve
D. medulla oblongata
E. two of the preceding

_____250. A patient involved in a sailing accident has lost the ability to sense the flow of air across his left arm and leg and to perceive pain on the right side of his body. These symptoms are suggestive of:
A. a hemisection of the right side of the brain stem above the level of the medulla
B. a hemisection of the left side of the spinal cord at the cervical level
C. a hemisection of the right side of the spinal cord at the thoracic level
D. a hemisection of the left side of the spinal cord at the thoracic level
E. none of the preceding

_____251. Severe damage to this area of the brain may result in cessation of breathing. It is the:
A. cerebrum
B. cerebellum
C. medulla oblongata
D. thalamus
E. hypothalamus

_____252. Gamma motor neurons:
A. are found in the dorsal root of a spinal nerve
B. stimulate intrafusal fibers
C. inhibit extrafusal fibers
D. are controlled by extrapyramidal tracts
E. two of the preceding

_____253. Muscles of facial expression are controlled by motor fibers in the _____ cranial nerves.
- A. first
- B. third
- C. fifth
- D. seventh
- E. ninth

_____254. The ability to discriminate between two mechanical stimuli applied to skin requires normal functioning of the:
- A. anterior spinothalamic tracts
- B. ventral spinocerebellar tracts
- C. dorsal columns
- D. parietal cortex of cerebrum
- E. two of the preceding

_____255. Transection (completely cutting through) of the anterior roots of the spinal nerves that supply the muscles of the posterior thigh would result in:
- A. inability to voluntarily extend the leg
- B. inability to elicit the knee-jerk reflex
- C. spastic paralysis of posterior thigh muscles
- D. flaccid paralysis of posterior thigh muscles
- E. two of the preceding

_____256. The autonomic receptors on the postganglionic neuron are:
- A. muscarinic
- B. adrenergic
- C. nicotinic
- D. sensitive to epinephrine
- E. none of the preceding

_____257. Which of the following is not one of the pigments found in cones?
- A. yellow   C. red
- B. blue   D. green

_____258. Discrimination of sound frequencies and intensities detected by the right ear involves:
- A. the right temporal cortex
- B. the left temporal cortex
- C. the frontal cortex
- D. the occipital cortex
- E. two of the preceding

_____259. Which of the following is (are) dominated by the left cerebral hemisphere?
   A.   artistic creativity   E.   A & D
   B.   intuitive thought
   C.   subtle meanings of language
   D.   logical thought as in science

_____260. Nicotinic receptors:
   A.   are adrenergic receptors
   B.   can be found on both sympathetic and parasympathetic neurons
   C.   respond to muscarine
   D.   are the same as curariform receptors
   E.   are designated as alpha or beta receptors

_____261. A lesion has blocked all transmission into and through the right half of the spinal cord at the 12th thoracic segment. The patient loses:
   A.   touch and pressure sensation from left leg
   B.   motor control of the right arm
   C.   pain and temperature sensation from left leg
   D.   motor control of the left arm
   E.   two of the preceding

_____262. A patient can see a stop sign, describe its shape and color, but is unable to interpret the meaning or significance of the sign. Assuming the patient has had a stroke (CVA) the area of probable damage would be the:
   A.   frontal lobe cortex
   B.   temporal-occipital association cortex
   C.   brainstem
   D.   thalamus
   E.   post-central gyrus

_____263. A Pacinian corpuscle is most responsive to:
   A.   pressure
   B.   chemical stimuli
   C.   temperature change
   D.   sound waves
   E.   light waves

_____264. The retina contains:
   A.   ganglion cells
   B.   photoreceptors
   C.   bipolar cells
   D.   horizontal cells
   E.   all of the above

_____265. Damage to the right anterolateral pathways at spinal segment C_3 could result in:
- A. loss of pain sensation from the right arm
- B. loss of temperature sensation from the left leg
- C. loss of pain sensation from the right leg
- D. loss of left patellar reflex (knee-jerk)
- E. two of the preceding

_____266. Alpha motorneurons:
- A. are found in the posterior root of a spinal nerve
- B. are controlled by upper motorneurons
- C. are lower motorneurons
- D. adjust sensitivity of the neuromuscular spindle
- E. two of the preceding

_____267. Parasympathetic stimulation:
- A. increases gastrointestinal motility
- B. increases pupil diameter
- C. increases heart rate
- D. increases airway diameter
- E. increases production of saliva that is very thick causing dry cotton mouth feeling

_____268. The patient has suffered a stab wound to the neck resulting in damage to the spinal cord as diagrammed below. The patient would be unable to:

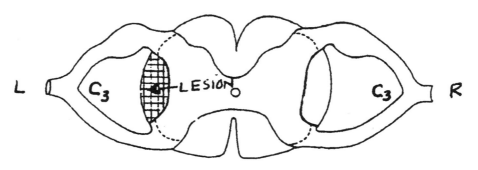

- A. sense cutaneous pain from the right leg
- B. voluntarily contract muscles of the right leg
- C. sense cutaneous temperature changes from the left leg
- D. demonstrate myotatic reflexes of the right leg
- E. do two of the above

_____269. Injury to the hypothalamus would most likely result in:
- A. inability to move the eyes
- B. loss of memory
- C. loss of body temperature control
- D. loss of the ability to perceive pain
- E. paralysis of the legs

431

270. Hypermetropia:
  A. occurs when light is focused in front of the retina
  B. is another name for presbyopia
  C. can be corrected by placing a convex lens in front of the eye
  D. means vision which is extraordinarily good
  E. none of the preceding

271. A time when you would want to simultaneously increase spindle sensitivity in both flexors and extensors at a given joint.
  A. when walking
  B. when running
  C. when swimming
  D. when standing at attention
  E. there is never such a time

272. Repeated exposure to the high intensity sounds of hard rock music (noise) has left you with an inability to hear sound frequencies at and near 20,000 Hz. A likely site of damage to the auditory system is the:
  A. organ of Corti, apical cochlea
  B. semicircular canal, superior
  C. saccule
  D. organ of Corti, basal cochlea
  E. spiral ganglion

273. Damage to the cranial nerve #1 would most likely interfere with the sense of:
  A. balance
  B. sight
  C. smell
  D. hearing
  E. taste

274. So that the tympanic membrane can vibrate normally, the _____ allows air pressure to become equal on either side.
  A. pinna
  B. cochlea
  C. eustachian tube
  D. semicircular canal
  E. utricle

275. As an object in your visual field is moved closer to your eyes, the visual image of the object remains clear because:
  A. the ciliary muscle relaxes
  B. your eyes diverge
  C. you have passed the near point
  D. the ciliary muscle contracts
  E. the pupils constrict

276. The withdrawal reflex:
  A. is monosynaptic
  B. is a brainstem reflex
  C. is primarily an ipsilateral
  D. requires input from the cerebral cortex
  E. involves ipsilateral stimulation of extensors

277. Which of the following is not an effector for the ANS?
  A. smooth muscle cell
  B. secretory cell of the salivary glands
  C. cardiac muscle cell
  D. pacemaker cell of the heart
  E. skeletal muscle cell

278. Reflexes of the brainstem or spinal cord:
  A. are always multisynaptic
  B. can always be voluntarily inhibited
  C. are always excitatory, never inhibitory
  D. may be ipsilateral or contralateral
  E. always require skeletal muscle as an effector

279. The anterior root of spinal nerve T$_{10}$ on the right side has been severed. Loss would include:
  A. cutaneous pain sensation from the right leg
  B. cutaneous temperature sensation from the left leg
  C. loss of voluntary activation of right leg skeletal muscles
  D. cutaneous touch and pressure from the area supplied by RT$_{10}$.
  E. flaccid paralysis of muscles supplied by RT$_{10}$.

280. The sensory information carried by fibers in the optic tract on the right side of the brain is derived from:
  A. the entire right visual field
  B. the nasal component of the left and right visual fields
  C. the temporal component of the left and right visual fields
  D. the nasal component of the right visual field and the temporal component of the left visual field
  E. the nasal component of the left visual field and the temporal component of the right visual field

281. The visual defect corrected by a concave lens that occurs because the image is focused in front of the retina is:
  A. myopia   D. emmetropia
  B. hypermetropia   E. cyclopia
  C. astigmatism

_____282. General parasympathetic stimulation results in:
- A. dilation of the pupil
- B. a decrease in heart rate
- C. secretion of thick, scant saliva
- D. a decrease in pancreatic enzyme secretion
- E. difficulty in swallowing

_____283. Angular acceleration of the body is detected by receptors in the:
- A. utricle
- B. saccule
- C. vestibule
- D. semicircular canals
- E. cochlea

_____284. A lesion in the right primary auditory cortex could result in:
- A. complete right ear deafness
- B. complete left ear deafness
- C. partial right ear deafness
- D. partial left ear deafness
- E. answers (C) and (D)

_____285. Which of the following nerves contain preganglionic parasympathetic fibers?
- A. tenth thoracic spinal nerve
- B. trigeminal nerve
- C. eighth cervical nerve
- D. facial nerve
- E. optic nerve

_____286. Which of the following reflects increased parasympathetic activity?
- A. increased heart rate
- B. increased blood flow to gut
- C. decreased pupillary diameter
- D. increased airway diameter
- E. two of the preceding

_____287. A near-sighted (myopic) individual:
- A. has a shortened eyeball
- B. use a biconcave lens for correction
- C. has a visual acuity of 1.00
- D. has the image of an object focused behind the fovea
- E. two of the preceding

_____288. Obstinence, irreverence, changes in motor behavior, overreaction to inconsequential stimuli, excessive anger, rage, and other disturbances of personality suggest possible damage to the:

A. temporal association cortex   D. mesencephalon
B. medulla oblongata   E. spinal cord
C. prefrontal cortex

_____289. Regardless of the modality of a stimulus, the sensation perceived when the receptor is stimulated is always the same. This is known as the:

A. law of adequate stimulus
B. all or none law
C. law of specific nerve energies
D. Murphy's law
E. Starling's law

_____290. Transection (completely cutting through) of the posterior roots of the spinal nerves that supply the muscles of the anterior thigh would result in:

A. inability to voluntarily extend the leg
B. inability to elicit the knee-jerk (patellar) reflex
C. spastic paralysis of anterior thigh muscles
D. flaccid paralysis of the anterior thigh muscles
E. two of the preceding

_____291. The eustachian tube functions:

A. to equalize air pressure on both sides of the tympanic membrane
B. to help maintain equilibrium
C. to magnify sounds
D. to protect sounds
E. all of the preceding

_____292. Involuntary control of heart rate, blood pressure, respiratory rate and the depth of breathing are functions of the:

A. seventh cranial nerves
B. medulla oblongata
C. eighth cranial nerves
D. eleventh cranial nerves
E. two of the preceding

_____293. Cones are:

A. receptors for color vision
B. found only in the optic disc
C. receptors for night vision
D. found primarily in the periphery of the retina
E. two of the preceding

_____294.   The gamma motorneurons of an extensor muscle:
A.   are inhibited when the alpha motorneurons of the opposing flexor are stimulated to produce flexion
B.   increase firing frequency as the opposing flexor muscle contracts.
C.   form motor units within the muscle
D.   are voluntarily activated by the temporal cortex of the cerebrum
E.   two of the preceding

_____295.   In the somatic motor system, intended movement is compared with actual movement and motor output adjusted to accomplish coordinated activity. The comparator function is accomplished by the:
A.   cerebrum
B.   cerebellum
C.   pons
D.   mesencephalon
E.   thalamus

_____296.   A nearsighted individual:
A.   has a shortened eyeball
B.   use a biconcave lens for correction
C.   has the image of an object focused behind the fovea
D.   both A and B
E.   A, B, and C

_____297.   Beta adrenergic receptors:
A.   are receptors for muscarine
B.   are nicotinic receptors
C.   can be blocked by curare
D.   are receptors for only norepinephrine
E.   increase heart rate when engaged with transmitter

_____298.   Based upon the changes in the frequency and amplitude of the brain's electrical activity there are _____ states of sleep.
A.   2
B.   3
C.   4
D.   5
E.   6

_____299.   Which of the following reflects increased sympathetic activity?
A.   decreased heart rate
B.   increased airway diameter
C.   vasodilation in the gut
D.   pupillary constriction
E.   decreased skeletal muscle blood flow

_____300. If one of the optic tracts were cut, the resulting blindness would involve:
   A.   only one eye
   B.   right and left thalamus
   C.   one half the visual field of both eyes
   D.   both occipital lobes of the cerebrum
   E.   none of the preceding

_____301. Discrimination of sound frequencies and intensities detected by the right ear involves:
   A.   the right temporal cortex
   B.   the left temporal cortex
   C.   the frontal cortex
   D.   the occipital cortex
   E.   two of the preceding

_____302. A person has suffered a stroke resulting in permanent damage to the area of the cerebral cortex indicated below. As a result of the damage, the person would:

LESION

   A.   be blind in one eye
   B.   be unable to hear in either ear
   C.   be completely paralyzed on the left side
   D.   have difficulty in differentiating cutaneous pain from cutaneous pressure stimuli
   E.   be unable to speak

_____303. The intensity of a pressure stimulus is sensed as a result of:
   A.   the frequency of action potentials generated in a single fiber
   B.   the number of fibers activated
   C.   the strength of each individual impulse
   D.   the speed of conduction of each individual impulse
   E.   both A and B

_____304. The trichromatic theory of color vision is based on:
   A.   red, yellow, green cones
   B.   red cones, green cones, rods
   C.   rods and cones
   D.   blue, red, and green cones
   E.   purple, orange, and pink cones

437

_____305. Conductive hearing loss results from damage to:
- A. outer ear
- B. middle ear
- C. inner ear
- D. cerebral cortex
- E. two of the above

_____306. The ability to bring the tips of your right and left index finger together with your eyes closed involves:
- A. the tactile sense
- B. the use of proprioreceptors
- C. nociceptors
- D. tonic discharge
- E. all of the above

_____307. Which tract is used to initiate rapid, fine-skilled movement?
- A. lateral corticospinal
- B. anterior spinothalamic
- C. lateral spinothalamic
- D. posterior column
- E. spinocerebellar

_____308. Subconscious coordination and control of skeletal muscles as when maintaining balance requires normal function of the:
- A. parietal cortex of the cerebrum
- B. cerebellum
- C. cochlea
- D. otolith organs of the utricle and saccule
- E. two of the preceding

_____309. General sympathetic stimulation results in:
- A. a decrease in heart rate
- B. an increase in gastrointestinal motility
- C. dilation of the pupil
- D. increased watery saliva secretion
- E. temporary paralysis of skeletal muscle

_____310. Transection (completely cutting through) of the anterior roots of the spinal nerves that supply the muscles of the anterior thigh would result in:
- A. inability to voluntarily extend the leg
- B. inability to elicit the knee-jerk (patellar) reflex
- C. spastic paralysis of anterior thigh muscles
- D. flaccid paralysis of the posterior thigh muscles
- E. two of the preceding

_____311.   As a person ages a loss of near vision occurs as the lens gradually loses its elasticity and the near point of accommodation recedes. This is know as:
A.   myopia
B.   presbyopia
C.   emmetropia
D.   hypermetropia
E.   glaucoma

_____312.   Contraction of the ciliary muscle causes:
A.   the eyeball to rotate laterally
B.   the pupil to become smaller
C.   the lens to become thinner (flatten)
D.   the lens to become thicker (bulge)
E.   the near-point to recede

_____313.   The anterior root of the spinal nerve T_8 has been severed. In the area of the body supplied by T_8 there would be loss of:
A.   cutaneous pain sensation
B.   cutaneous light touch sensation
C.   conscious sense of joint movement
D.   voluntary control of skeletal muscle and reflex control
E.   reflex control of skeletal muscle but not voluntary control

_____314.   Otoliths (calcium carbonate) play a role in the function of:
A.   the crista
B.   the macula
C.   the organ of Corti
D.   the tympanic membrane
E.   none of the preceding

_____315.   Which of the following would you expect to observe in a patient with a tumor of the cerebellum?
A.   loss of vision
B.   loss of hearing
C.   inability to perceive pain
D.   inability to execute any voluntary movements
E.   inability to execute smooth, steady movements

_____316.   You accidentally step on a tack with your bare foot and reflexively withdraw your foot. The:
A.   contralateral flexors of the knee were stimulated to contract
B.   ipsilateral extensors of the knee were stimulated to contract
C.   the contralateral extensors of the knee were inhibited
D.   the ipsilateral flexors of the knees were stimulated to contract
E.   two of the preceding

_____317.   Rotation of the body around a vertical axis, as in spinning on iceskates, is detected by:
    A.   otoliths of the saccule
    B.   hair cells in the ampullae of semicircular canals
    C.   hair cells in the organ of Corti
    D.   the cochlear division of cranial nerve VIII
    E.   maculae of the vestibules

_____318.   Stimulation of the eye's photoreceptors by a pressure stimulus results in the perception of a visual sensation rather than a pressure sensation. This demonstrates the law of:
    A.   adequate stimulus
    B.   all or none
    C.   specific nerve energies
    D.   averages
    E.   Murphy

_____319.   When the ciliary muscle contracts:
    A.   the lens becomes thinner
    B.   the pupil becomes smaller
    C.   the pupil becomes larger
    D.   the lens becomes thicker
    E.   the eyeball moves laterally

_____320.   What is contained in the medulla oblongata?
    A.   centers that control core body temperature
    B.   corpus callosum
    C.   primary cardiovascular control center
    D.   superior and inferior colliculi
    E.   nuclei for cranial nerves 2, 3, and 4

MATCHING. CHOOSE THE BEST ANSWER. SOME ANSWERS MAY BE USED MORE THAN ONCE OR NOT AT ALL.

A. sympathetic characteristic or effect
B. parasympathetic characteristic or effect
C. both A and B
D. neither A nor B

_____321. contraction of papillary dilator muscle

_____322. increased secretion of gastric juices

_____323. dilation of blood vessels that supply skeletal muscle

_____324. contraction of skeletal muscle

_____325. decreased rate of contraction of cardiac muscle

A. facial nerve
B. trigeminal nerve
C. oculomotor nerve
D. trochlear nerve
E. none of the preceding

_____326. monitors blood pressure in the carotid artery

_____327. controls the ciliary muscle

_____328. transmits pain information from the face

_____329. moves the eye to direct the pupil medially and downward

_____330. sensory fibers convey taste information from the posterior 1/3 of the tongue

_____331. motor fibers supply muscles that raise the eyebrows

_____332. motor fibers supply muscles that clench the teeth

_____333. contains parasympathetic preganglionic fibers that supply the pulillary constrictor muscle

A. frontal lobe cerebral cortex
B. parietal lobe cerebral cortex
C. temporal lobe cerebral cortex
D. occipital lobe cerebral cortex
E. none of the preceding

_____334. planning, anticipating what needs to be done, sorting trivial from important, controlling motor related behaviors

_____335. feeling the pain of stubbing your toe

_____336. uses sensory information from the inner ear to coordinate/control skeletal muscles so as to maintain balance or posture without conscious effort

_____337. stores information about the melody of a tune you're going to whistle

_____338. differentiates between red and purple, blue and green, light and dark

_____339. recognition of the difference between a C# played on a piano vs a C# played on an organ

_____340. reflexively withdrawing the foot from a sharp object that was stepped on

_____341. visualizing with the eyes closed

A. medulla oblongata
B. pons
C. midbrain
D. hypothalamus
E. thalamus

_____342. receives and processes visual information from the optic tracts before sending to the visual cortex

_____343. controls core body temperature via the endocrine and autonomic nervous system

_____344. contains nuclei that control muscles of facial expression

_____345. speeds up the heart and slows breathing rate

A.   right lateral corticospinal tract
B.   left lateral spinothalamic tract
C.   right anterior spinothalamic tract
D.   left anterior corticospinal tract
E.   none of the preceding

_____346.   damage to this tract at T_{10} could result in loss of pain sensation from the skin of the left thigh

_____347.   this tract conveys light touch and pressure information from the right arm

_____348.   severing all fibers of this tract causes major paralysis on the left side of the body

_____349.   this tract gives the brain information about skeletal muscle contraction, joint movement, and vibration on the left side of the body.

_____350.   injury to this tract at C_4 could result in the loss of voluntary control of about 85% of the motor units on the left side of the body from the neck down

A.   facial nerve
B.   trigeminal nerve
C.   optic nerve
D.   hypoglossal nerve
E.   none of the preceding

_____351.   damage to this nerve on the right side causes the tongue to deviate to the right

_____352.   contains papillary constrictor nerve fibers

_____353.   Bell's palsy

_____354.   supplies muscles of mastication (clench teeth)

_____355.   smile

A. temporal lobe cortex
B. occipital lobe cortex
C. frontal lobe cortex
D. parietal lobe cortex
E. none of the preceding

_____356. planning motor activity, ignoring trivia

_____357. distinguishing frequencies and intensities of muscle notes

_____358. distinguishing red and green from orange and yellow

_____359. identifying the localizing pain from a burned finger

_____360. initiating rapid, fine-skilled movements

A. thalamus
B. medulla oblongata
C. hypothalamus
D. mesencephalon
E. none of the preceding

_____361. integrates and relays somatic and special sensory information

_____362. contains centers controlling thirst and appetite

_____363. if this part of the brain is damaged, breathing could stop

_____364. nuclei in this part of the brain are involved with visual reflexes such as accommodation

_____365. damage to nuclei located here could result in a loss of peripheral vision

A. right posterior white matter
B. left anterior root
C. left lateral white matter
D. right posterior root
E. none of the preceding

_____366. a lesion here could result in spastic paralysis involving 85% of the ipsilateral motor units below the level of the lesion

_____367. an appreciation of a vibrating tuning fork placed on the left knee could be lost if this area of the spinal cord at T_8 were destroyed

_____368. transection of this spinal nerve root results in ipsilateral flaccid paralysis

_____369. transection of this spinal nerve root results in contralateral anesthesia

_____370. damage here at C_2 could result in a loss of temperature sensation from the skin on the right forearm

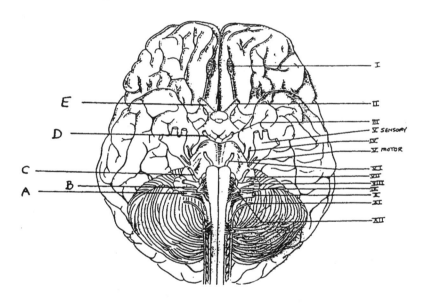

_____371. contains nasal and temporal fibers from retina

_____372. relays taste information from posterior 1/3 of the tongue

_____373. stimulates salivary glands, relays taste from anterior tongue

_____374. contains motor fibers of pupillary constrictor muscles

_____375. monitors blood pressure from thoracic blood vessels

445

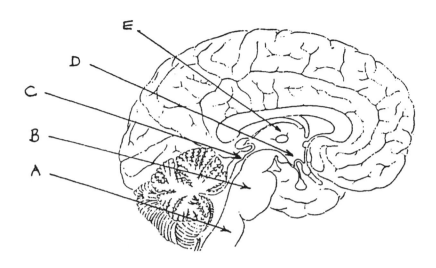

_____376. contains heat loss and heat gain centers

_____377. contains primary cardiac control centers

_____378. crudely integrates somatic and special sensory information

_____379. cochlear nuclei are mostly in this part of the brain

_____380. visual reflexes are centered here

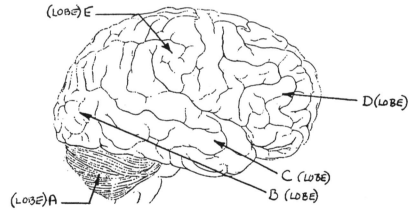

_____381. remembering favorite songs

_____382. distinguishing pink from purple

_____383. suppressing the urge to become enraged

_____384. localizing and feeling the pain of a sprained ankle

_____385. coordination and subconscious control of skeletal muscles as when maintaining balance during walking

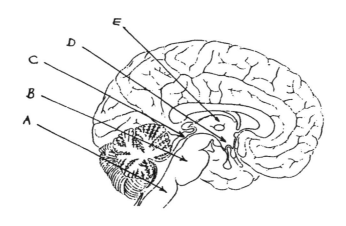

_____386.   contains centers that regulate ECF osmotic pressure

_____387.   contains primary respiratory control centers

_____388.   contains nuclei that integrate and relay optic tract signals

_____389.   contains nuclei that relay somatic sensory information from the face

_____390.   contains nuclei of the accommodation reflex

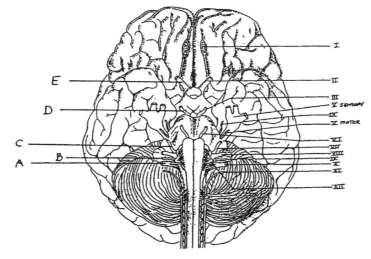

_____391.   cut this nerve and complete blindness in one eye occurs

_____392.   monitors blood pressure and blood chemistry in carotid artery

_____393.   motor fibers control muscles of facial expression

_____394.   contains motor fibers of the accommodation reflex

_____395.   motor fibers cause the heart to decrease its rate of beating

PHYSIOLOGY OUTLINES 13-18

MULTIPLE CHOICE. SELECT THE BEST ANSWER.

_____396.　At the beginning of inspiration, which force favors inflation of the lungs?
　　A.　alveolar surface tension
　　B.　elastic recoil of the lungs
　　C.　elastic recoil of the thorax
　　D.　airway resistance
　　E.　intrapulmonic pressure

_____397.　The distance between successive R waves on Lead II EKG is 25 mm. The heart rate is:
　　A.　60 BPM
　　B.　70 BPM
　　C.　80 BPM
　　D.　120 BPM
　　E.　75 BPM

_____398.　In a 20 ml sample of whole blood, packed red cell volume is 9 ml. The hematocrit is therefore:
　　A.　1190　　　　　　D.　45%
　　B.　4.5 ml　　　　　E.　45 ml
　　C.　11 ml

_____399.　Functional residual capacity is the sum of:
　　A.　ERV + TV
　　B.　ERV + IRV
　　C.　RV + ERV
　　D.　TV + RV
　　E.　IRV + TV

_____400.　The mitral and aortic valves simultaneously open during:
　　A.　rapid ejection
　　B.　isovolumetric contraction
　　C.　isovolumetric relaxation
　　D.　diastasis
　　E.　no part of the cardiac cycle

_____401. Which volume remains constant throughout a person's life?
  A.   vital capacity
  B.   expiratory reserve volume
  C.   inspiratory reserve volume
  D.   residual volume
  E.   none of the preceding

_____402. Ventilation/Perfusion ratios in the standing person are typically high, compared to the ideal, in the apex of the lung because:
  A.   ventilation of the apex is better than most other areas of the lung
  B.   gravity and vascular reflexes significantly reduce blood flow to the apex
  C.   the greatest changes in the intrapulmonic pressure occur in the apex
  D.   the clavicle (collarbone) interferes with elevation of the first rib
  E.   pulmonary vascular resistance is less than systemic vascular resistance

_____403. All four intracardiac valves are closed during:
  A.   no part of the cardiac cycle
  B.   isovolumetric contraction
  C.   rapid filling of the ventricles
  D.   rapid ejection
  E.   AV node depolarization

_____404. A lengthening of the RR interval suggests:
  A.   tachycardia
  B.   bradycardia
  C.   a slowing of pacemaker signal conduction to the ventricles
  D.   SA node damage
  E.   2:1 heart block

_____405. Given the following, compute cardiac output: H.R. = 70 BPM, EDV = 80 ml, ESV = 15 ml:
  A.   6650 ml
  B.   4.5 L
  C.   5.6 L/min
  D.   4550 ml/beat
  E.   4550 ml/min

_____406. Which factor from Poiseuille's law inversely affects blood flow (volume per unit of time)?
  A.   hydrostatic pressure difference
  B.   blood vessel radius
  C.   blood viscosity
  D.   cross-sectional area of the blood vessel
  E.   molecular weights of plasma proteins

_____407. Given the following data, compute alveolar ventilation:
ADS = 150 ml
VC = 4800 ml
RR   = 15 CPM
IRV = 3300 ml
ERV = 1000 ml

A.   2250 ml/min
B.   5.25 L/min
C.   7.2 L/min
D.   4500 ml/min
E.   5250 ml

_____408. Numerically speaking, if pulmonary blood flow (from RV to LA) were 5 units and pulmonary vascular resistance (from RV to LA) were 6 units, then the difference between mean pressure at the beginning of the pulmonary trunk and the mean pressure at the ends of pulmonary veins would be: (remember $F = P/R$)
A.   1 unit
B.   30 units
C.   62 units
D.   5 units
E.   6 units

_____409. If pulse pressure is 90 torr and mean arterial pressure is 90 torr, diastolic pressure is approximately:
A.   180 torr
B.   zero
C.   120 torr
D.   60 torr
E.   80 torr

_____410. Given the following data, compute capillary hydrostatic pressure: capillary filtration pressure = 10 torr, plasma protein osmotic pressure = 25 torr, tissue hydrostatic pressure = 5 torr, tissue protein osmotic pressure = 5 torr.
A.   15 torr   D.   35 torr
B.   25 torr   E.   40 torr
C.   30 torr

_____411. According to the carotid baroreceptor reflex:
A.   an increase in carotid arterial P_{CO_2} increases cardiac output
B.   a decrease in carotid arterial PO_2 increases cardiac output
C.   a decrease in carotid arterial pH increases cardiac output
D.   a decrease in carotid pressure decreases cardiac output
E.   none

_____412. The maximal volume of gas that can be expired after a resting tidal expiration is called:
  A. expiratory reserve volume
  B. expiratory capacity
  C. functional residual capacity
  D. vital capacity
  E. inspiratory reserve volume

_____413. The effectiveness of both antibody-mediated immune responses and cell-mediated immune responses to the presence of an antigen is most dependent on:
  A. eosinophils
  B. CD4+lymphocytes (T4 cells)
  C. CD8+lymphocytes (T8 cells)
  D. plasma cells
  E. basophils

_____414. During blood coagulation, a soluble plasma protein produced by the liver is converted to an insoluble plasma protein. The conversion is the main event of:
  A. Stage I   D. fibrinolysis
  B. Stage II   E. kwashiorkor
  C. Stage III

_____415. On destruction of the red blood cells and metabolism of hemoglobin:
  A. all components of Hb are conserved except iron which is excreted
  B. all components of the molecule are conserved
  C. all components of the molecule are excreted
  D. iron and protein (amino acids) are conserved and the remainder is excreted as waste in the urine or feces
  E. the globin is conserved and the heme is excreted

_____416. Which of the following is the largest?
  A. inspiratory reserve volume
  B. tidal volume
  C. expiratory reserve volume
  D. residual volume
  E. inspiratory capacity

_____417. When a greater change in interpleural pressure is required to effect a former change in intrapulmonic volume:
  A. the lungs have become more compliant
  B. compliance of the thorax has increased
  C. pulmonary compliance has decreased
  D. airway resistance has decreased
  E. pneumonthorax has occurred

451

_____418.   According to the oxyhemoglobin dissociation curve:
A.   an increase in blood pH (more alkaline) favors the unloading of oxygen from hemoglobin
B.   when the body is at rest, less than one-third of available hemoglobin's oxygen-carrying capacity is used to deliver oxygen to tissues
C.   when breathing air, hemoglobin saturation in pulmonary arterial blood is 98%
D.   hemoglobin can carry carbon dioxide in addition to oxygen
E.   the hemoglobin content of normal blood is 14

_____419.   An increase in blood pressure within the venae cavae and right atrium signifies an increase in venous return and reflexly:
A.   excites the cardioinhibitory center
B.   depresses cardioaccelerator center activity
C.   increases heart rate
D.   lowers systemic arterial pressure
E.   causes drowsiness

_____420.   Which of the following is true concerning regulation of alveolar ventilation:
A.   the regulation of alveolar ventilation is a function of primary respiratory control centers located in the Pons
B.   the expiratory center normally does not play an active role (stimulating muscles) during resting respiration
C.   an increase in PO_2 in systemic arterial blood is the most powerful stimulus to the respiratory centers
D.   the respiratory centers tend to be stimulated by an increase in systemic arterial pH
E.   AV = (TV)(RR)

_____421.   Given the following data, compute the ejection fraction:
CO = 5.6 L/min   HR = 80 bpm   SV = 70 ml
ESV = 50 ml   EDV = 120 ml

A.   70 ml   C.   70%   E.   1.25%
B.   58 ml   D.   58%

_____422.   Given the following data, compute capillary absorption pressure: H.B.P. = 30 torr; P.P.O.P. = 28 torr; THP = 4 torr; TPOP = 2 torr.
A.   8 torr
B.   5 torr
C.   3 torr
D.   1 torr
E.   Zero

_____423. The appropriate blood volume of a 70 kgm adult is:
A. 3 liters   D. 12 liters
B. 5 liters   E. 15 liters
C. 9 liters

_____424. If hemoglobin content is 13 grams per deciliter and the combining power of hemoglobin for oxygen is 1-1/3 ml of oxygen per gram of hemoglobin, then the oxygen-carrying capacity of the blood:
A. cannot be computed without more data
B. is approximately 20 vol. percent
C. is approximately 15 ml of O_2 per liter
D. is 17-18 ml of O_2 per deciliter of blood
E. is 17.4 gms/ml

_____425. Which of the following can safely be infused into a B⁻ recipient without *agglutination* and without sensitizing the recipient to the Rh antigen?
A. plasma of whole blood type AB⁺
B. whole blood type B⁺
C. whole blood type A⁻
D. serum of whole blood type O⁺
E. two of the preceding

_____426. The period of time in the cardiac cycle that immediately precedes the rapid ejection phase of ventricular systole is a very short phase called:
A. ventricular diastole
B. atrial systole
C. isovolumetric contraction
D. isovolumetric relaxation
E. diastasis

_____427. One of the following is abnormally high. Which one?
A. WBC: 8000 per microliter
B. RBC: 5.3×10^6 per microliter
C. ESR: 6.0 mm/hr
D. Hematocrit: 60 percent
E. MCH: 30 μμg (micromicrograms)

_____428. Which of the following is a normal value for MCHC?
A. 90%
B. 34%
C. 45%
D. 27 pg
E. 30 μμg

429. Given the following, compute capillary absorption pressure: hydrostatic blood pressure = 20 torr; plasma protein osmotic pressure = 25 torr; tissue hydrostatic pressure = 5 torr; tissue protein osmotic pressure = 5 torr.
   A. 5 torr
   B. 10 torr
   C. 15 torr
   D. 20 torr
   E. 25 torr

430. As regards the systemic capillary, which of the following favors increased filtration and possibly the development of peripheral edema?
   A. increased plasma colloidal osmotic pressure (PPOP)
   B. decreased hydrostatic blood pressure (HBP)
   C. increased interstitial hydrostatic pressure (THP)
   D. increased interstitial colloidal osmotic pressure (TPOP)
   E. none of the above

431. Which of the following may be infused into an A⁻ recipient without major or minor agglutination occurring?
   A. whole blood type B⁺
   B. serum from type O⁻ blood
   C. serum from type A⁺ blood
   D. whole blood type AB⁻
   E. whole blood type O⁻

432. Upon death of the erythrocyte, the hemoglobin contained within the cell:
   A. is completely destroyed and excreted into the intestine
   B. remains intact and is incorporated into new erythrocytes
   C. is broken down into heme and globin from which iron and amino acids are extracted and conserved
   D. remains intact and is excreted into bile
   E. is released into the plasma and returned to the bone marrow for reuse

433. According to the oxyhemoglobin dissociation curve the change in the saturation of hemoglobin that occurs at rest when systemic arterial blood gives up oxygen in the systemic capillary is approximately:
   A. 0-10 percent
   B. 10-20 percent
   C. 20-30 percent
   D. 70 percent
   E. 100 percent

434. Pulmonary compliance:
  A. decreases as the lungs become maximally inflated
  B. would not be affected by pneumothorax
  C. does not change during maximal inspiration
  D. increases as the lungs become maximally inflated
  E. two of the preceding

435. Given the following data, compute ESV: HR = 80 beats/min.; CO = 4.8 L/min; EDV = 80 ml
  A. 10 ml
  B. 20 ml
  C. 30 ml
  D. 60 ml
  E. 80 ml

436. Which condition would tend to increase blood viscosity above normal?
  A. pernicious anemia
  B. leukopenia
  C. polycythemia vera
  D. hypochromic anemia
  E. two of the preceding

437. The volume of air remaining in the lungs at the end of a tidal (resting) expiration is:
  A. tidal volume
  B. expiratory capacity
  C. residual volume
  D. functional residual capacity
  E. expiratory reserve volume

438. Which of the following values for platelet count would be indicative of thrombocytopenia?
  A. 20,000 per μl
  B. 200,000 per μl
  C. 2 million/ml
  D. 2×10^5 per μl
  E. all of the preceding

439. A decrease in pulmonary compliance means:
  A. the lungs expand with less effort during inspiration
  B. a greater volume change occurs for a given change in interpleural pressure
  C. a greater expenditure of energy is required to inflate the lungs
  D. the person has asthma
  E. two of the preceding

_____440.   Which of the following is not a function of the upper division of the respiratory system?
  A.   humidification of inspired air
  B.   warming of inspired air
  C.   cleansing of inspired air
  D.   gas exchange with the blood
  E.   olfaction

_____441.   End-diastolic volume (EDV):
  A.   minus stroke volume equals ESV
  B.   is inversely proportional to ventricular compliance
  C.   cannot vary
  D.   is directly proportional to heart rate
  E.   increases as atrial pressure decreases

_____442.   Given the following data, compute heart rate:
          SV = 60 ml/beat
          EDV = 70 ml
          ESV = 10 ml
          CO = 5.4 L/min
          HR = _____

  A.   60 BPM
  B.   70 BPM
  C.   80 BPM
  D.   90 BPM
  E.   cannot be computed without more data

_____443.   The fraction of the total blood volume occupied by erythrocytes is:
  A.   elevated in anemia
  B.   normally about 12-16% in females
  C.   depressed in polycythemia
  D.   termed the hematocrit
  E.   higher in females than in males

_____444.   Which of the following values for cell count (per ul) would indicate polycythemia vera?
  A.   neutrophils: 7500
  B.   lymphocytes: 3000
  C.   erythrocytes: 8.5 million
  D.   monocytes: 500
  E.   none of the preceding

_____445.   Which of the following values are normal for mean corpuscular hemoglobin?
  A.   10-20 µµg   C.   12-19 µµg   E.   45-50 µµg
  B.   13-19 µµg   D.   27-32 µµg

_____446. End-diastolic volume (EDV):
  A. minus stroke volume equals ESV
  B. is inversely proportional to ventricular compliance
  C. cannot vary
  D. is directly proportional to rate
  E. decreases as atrial pressure increases

_____447. The following leukocyte plays an important role in the destruction of immune complexes.
  A. monocyte
  B. neutrophil
  C. eosinophil
  D. basophil
  E. lymphocyte

_____448. Systolic blood pressure is 140 torr. Diastolic blood pressure is 70 torr. Mean arterial blood pressure is:
  A. 78.3 torr
  B. 88.3 torr
  C. 93.3 torr
  D. 108.3 torr
  E. 55 torr

_____449. The law of the heart (Starling) accounts for:
  A. the ventricles' ability to increase cardiac output as venous return increases
  B. the A-V node becoming pacemaker for the ventricles in case of S-A node damage
  C. the inability of cardiac muscle to become tetanized
  D. dilation of coronary arterioles during arterial hypoxia
  E. use of the heart as a symbol for St. Valentine

_____450. If a person's combined WBC were 7×10^3 cells per microliter, a normal monocyte count would be (in cells per microliter):
  A. 3500   D. 5
  B. 350   E. 1×10^3
  C. 35

_____451. One or more of the following combinations suggests hypochromic anemia. Which one?
  A. hemoglobin - 12 gms/dl, RBC 4.8 million/ul
  B. hemoglobin - 18 gms/dl, RBC 5.4 million/ul
  C. hemoglobin - 9 gms/dl, RBC 4.0 million/ul
  D. hemoglobin - 14 gms/dl, RBC 5.0 million/ul
  E. two of the preceding

_____452. One of the following statements is false. Which one?
    A.   resting cardiac output for an adult is about 5L/min
    B.   ventricular muscle has a long absolute refractory period
    C.   at no time during the cardiac cycle are both atria and ventricles in systole
    D.   at no time during the cardiac cycle are both atria and ventricles in diastole
    E.   in a normal heart the SA node and the AV node depolarize with the same frequency

_____453. Given the following data, determine alveolar ventilation: ADS: 100 ml; RR: 15 cpm; VC: 3500 ml; RV: 1200 ml; TV: 400 ml.
    A.   1300 ml
    B.   300 ml/min
    C.   5000 ml/min
    D.   3000 ml/min
    E.   4.5 L/min

_____454. During normal resting expiration:
    A.   intrapulmonic pressure falls below atmospheric pressure
    B.   intrapulmonic volume increases
    C.   intrapulmonic pressure decreases
    D.   intrapulmonic volume decreases
    E.   both answers (C) and (D) occur

_____455. Capillary filtration pressure will increase if the following pressure(s) decrease(s).
    A.   hydrostatic blood pressure (capillary)
    B.   plasma protein osmotic pressure (oncotic pressure)
    C.   tissue protein osmotic pressure
    D.   tissue hydrostatic pressure
    E.   two of the preceding

_____456. Cardiac output is 5.4 liters per minute and heart rate is 60 beats per minute; therefore, the difference between end diastolic volume and end systolic volume is:
    A.   486 ml
    B.   4.8 liters
    C.   6 liters
    D.   90 ml
    E.   60 ml

_____457. During inspiration (eupnea):
A. intrapulmonic pressure is *never* greater than atmospheric pressure
B. intrapleural pressure is *never* less than atmospheric pressure
C. intrapleural pressure is *never* less than intrapulmonic pressure
D. answers A and B are correct but C is incorrect
E. answers A, B, and C are correct

_____458. Which of the following contains antibodies against antigens A, B, and D?
A. serum of whole blood type O^+
B. whole blood type A^-
C. whole blood type B^+
D. plasma of whole blood type AB^+
E. none of the preceding

_____459. _____ cells produce and secrete antibodies.
A. T_4 helper
B. B-plasma
C. T_8 cytotoxic
D. T_4 inducer
E. T_8 suppressor

_____460. According to the carotid baroreceptor reflex:
A. an increase in carotid arterial P_{CO_2} increases cardiac output
B. a decrease in carotid arterial P_{O_2} increases cardiac output
C. a decrease in carotid arterial pH increases cardiac output
D. an increase in carotid pressure decreases cardiac output
E. all of the preceding

_____461. End-systolic volume (ESV):
A. increases when ventricular contractility increases
B. is always less than end-diastolic volume
C. is the amount of blood in the ventricle just before the ventricle contracts
D. is not affected by arterial pressure
E. two of the preceding

_____462. Which of the following is a renal hormone that stimulates hematopoiesis?
A. aldosterone
B. enterogastrone
C. erythropoietin
D. prothrombin
E. estrogen

_____463.   Given the following data, compute heart rate: PR interval = 0.16 sec, QT interval = 0.36 sec, RR interval = 0.60 sec.
A.   72 beats/min   C.   86 beats/min   E.   70 beats/min
B.   100 beats/min   D.   94 beats/min

_____464.   When airways become restricted, such as in asthma:
A.   ventilation/perfusion reverses from base to apex
B.   pulmonary compliance increases
C.   statis pulmonary compliance increases
D.   dynamic pulmonary compliance decreases
E.   two of the preceding

_____465.   When we inflate the lungs during a normal respiratory cycle, we have to work against all of the following forces except:
A.   elastic recoil of the lung
B.   elastic recoil of the chest wall
C.   surface tension in the lung
D.   airway resistance
E.   two of the preceding

_____466.   Which of the following account for most of the O_2 and CO_2 transport in the blood?
A.   oxyhemoglobin, dissolved CO_2
B.   reduced hemoglobin, carbaminohemoglobin
C.   bicarbonate, oxyhemoglobin
D.   oxyhemoglobin, carbaminohemoglobin
E.   dissolved O_2 and CO_2

_____467.   In pulmonary edema, pulmonary diffusion may decrease because:
A.   surface area of the pulmonary membrane increases
B.   thickness of the pulmonary membrane increases
C. CO_2 is more soluble than O_2 in body water
D. PO_2 decreases
E.   two of the preceding

_____468.   Given the following, compute capillary filtration pressure: hydrostatic blood pressure = 40 torr; plasma protein osmotic pressure = 20 torr; tissue hydrostatic pressure = 10 torr; tissue protein osmotic pressure = 10 torr.
A.   5 torr
B.   10 torr
C.   15 torr
D.   20 torr
E.   25 torr

_____469. Cytotoxic chemical production is the primary function of which cell?
  A. erythrocyte
  B. eosinophil
  C. T_8 cell (CD8+)
  D. T_4 cell (CD4+)
  E. B-lymphocyte

_____470. T_8 lymphocytes:
  A. secrete antibodies
  B. help T_4 lymphocytes kill infected macrophages
  C. do not recognize MHC proteins
  D. eventually suppress an immune response
  E. two of the preceding

_____471. If tidal volume is 600 ml, anatomical dead space is 150 ml, and alveolar ventilation is 9 liters/min, then respiratory rate is:
  A. 10 cycles/min
  B. 15 cycles/min
  C. 20 cycles/min
  D. 25 cycles/min
  E. 30 cycles/min

_____472. Given the following, compute cardiac output: ESV = 10 ml, EDV = 90 ml, HR = 80 BPM:
  A. 5 L/min   D. 6.4 L/min
  B. 7.2 L/min   E. cannot be computed using data given
  C. 5.6 L/min

_____473. The maximal volume of gas that can be expired after a maximal inspiration is called:
  A. expiratory reserve volume
  B. expiratory capacity
  C. functional residual capacity
  D. vital capacity
  E. inspiratory reserve volume

_____474. Which of the following can be safely infused into an AB⁻ recipient, even if the recipient has been previously sensitized to the Rh antigen, and no *agglutination* will occur?
  A. whole blood type A⁻
  B. whole blood type B⁺
  C. serum of whole blood type O⁺
  D. plasma of whole blood type AB⁺
  E. two of the preceding

_____475. The lymphocyte most responsible for initiating the body's secondary immune response is the :
   A. T_8 lymphocyte
   B.   plasma cell
   C. T_4 lymphocyte
   D.   memory cell
   E.   suppressor cell

_____476. Given the following data, compute heart rate: PR interval = 0.16 sec, QT interval = 0.36 sec, RR interval = 0.50 sec.
   A.   72 beats/min
   B.   100 beats/min
   C.   86 beats/min
   D.   94 beats/min
   E.   120 beats/min

_____477. If lymphatic drainage of the lower extremity were blocked, the extremity would become edematous and swell because:
   A.   capillary permeability decreases
   B.   proteins accumulate in interstitial spaces and pull fluid out of the blood
   C.   systemic blood pressure in the extremity decreases
   D.   the liver can't manufacture an adequate number of plasma proteins
   E.   capillary absorption pressure increases

_____478. When bacteria invade the body, which one of the following cells is most likely to suffer the highest mortality rate?
   A.   B lymphocytes
   B.   monocytes
   C.   eosinophils
   D.   basophils
   E.   neutrophils

_____479. Which of the following does not coagulate and contains antibodies against antigens A, B, and D?
   A.   serum of whole blood type O^+
   B.   plasma of whole blood type AB^+
   C.   serum of whole blood type AB^-
   D.   whole blood type O^-
   E.   none of the preceding

_____480. Which of the following bloods could safely be recipient for a donation of one pint of AB$^+$ serum?
A.   O$^+$
B.   A$^-$
C.   AB$^-$
D.   O$^-$
E.   all of the preceding

_____481. In pneumothorax and complete atelectasis involving the right lung:
A.   interpleural pressure would not be affected
B.   intrapulmonic pressure would become negative
C.   interpleural pressure would increase toward atmospheric
D.   perfusion of the right lung would increase
E.   perfusion of the right lung would not change

_____482. Blood flow through the pulmonary circuit (vol./time) equals blood flow through the systemic circuit (vol./time) even though P for the pulmonary circuit is nearly five times less than P for the systemic circuit. The reason is:
A.   right and left ventricles function together
B.   pulmonary resistance is nearly five times less than systemic resistance
C.   pulmonary vessels cannot constrict
D.   oxygen must be absorbed by pulmonary blood at the same rate its being removed from systemic blood
E.   not evident in any of the preceding choices

_____483. An abnormal lengthening of the PR interval suggests:
A.   tachycardia
B.   bradycardia
C.   a slowing of pacemaker signals conduction to the ventricles
D.   SA node damage
E.   2:1 heart block

_____484. On a standard electrocardiogram, the distance from the beginning of one P wave to the beginning of the next P wave is 20 millimeters. The recorded heart rate is therefore:
A.   60 BPM
B.   75 BPM
C.   85 BPM
D.   120 BPM
E.   100 BPM

_____485. Since SV = EDV-ESV, left ventricular SV will be increased by all of the following *except*:

    A.   increased atrial pressure

    B.   increased ventricular pressure

    C.   increased ventricular contractility

    D.   increased ventricular compliance, i.e., an increase in ease with which the ventricle can be stretched

    E.   increased aortic diastolic pressure

_____486. All of the following describe the cell-mediated immune response *except*:

    A.   is the responsibility of the macrophages and the T-lymphocytes

    B.   T-lymphocytes recognize free antigens in tissue fluid

    C.   helper T_4 cells regulate cytotoxic T_8 and suppressor T_8 cells

    D. T_4 cells induce T_8 lymphocytes to differentiate into cytotoxic T_8 and suppressor T_8 cells

    E.   suppressor T_8 cells function to turn off cytotoxic T_8 cells and helper T_4 cells

_____487. Phase II of blood coagulation requires Ca^{++} and:

    A.   fibrinogen   D.   prothrombin

    B.   thrombin   E.   heparin

    C.   fibrin

_____488. During the cardiac cycle the papillary muscles and chorda tendinae function to:

    A.   close semilunar valves

    B.   open atrioventricular valves

    C.   eject blood into arterial systems

    D.   prevent eversion of the AV valves

    E.   relay pacemaker signals to the ventricles

_____489. If alveolar ventilation is 4900 ml/min, anatomical dead space is 150 ml and respiratory rate is 14 cycles per minute, then tidal volume is:

    A.   475 cc

    B.   2100 cc

    C.   525 cc

    D.   707 cc

    E.   500 cc

_____490. Cardiac output is 5.4 liters per minute and heart rate is 90 beats per minute; therefore, stroke volume is:

    A.   486 ml

    B.   4.8 liters

    C.   6 liters

    D.   600 ml

    E.   60 ml

_____491.   Functions of this cell include local control of blood clotting and regulation of blood flow in an area of tissue damage. The cell is called a(n):
A.   neutrophil
B.   basophil
C.   eosinophil
D.   monocyte
E.   suppressor T cell

_____492.   Given the following, compute cardiac output: ESV = 10 ml, EDV = 80 ml, HR = 80 BPM.
A.   5 L/min
B.   7.2 L/min
C.   5.6 L/min
D.   6.4 L/min
E.   cannot be computed by using the given data

_____493.   Which cell secretes interleukins that stimulate B-cell development?
A. T_4 helper cell
B.   red blood cell
C. T_8 cytotoxic cell
D. T_8 suppressor cell
E.   liver macrophage

_____494.   Given the following data, determine alveolar ventilation: ADS: 100 ml; RR: 15 cpm; VC: 3500 ml; RV: 1200 ml; TV: 400 ml.
A.   1300 ml
B.   300 ml/min
C.   5000 ml/min
D.   3000 ml/min
E.   4.5 L/min

_____495.   Consider the relationship between flow (F), pressure (P), and resistance (R) to the flow of blood through a blood vessel and pick the correct statement from below:
A.   P = F/R
B.   R = PF
C.   if flow remains constant and resistance to flow increases, then pressure falls
D.   an increase in blood viscosity increases vascular resistance
E.   if pressure increases, flow will also increase unless resistance increases in proportion to pressure

496. Compared to the systemic circulation, the pulmonary circulation:
   A. has a higher pressure gradient from beginning to end
   B. offers less resistance to the flow of blood
   C. has a higher capillary filtration pressure
   D. has a lower rate of blood flow through the circuit
   E. two of the preceding

497. Numerically, which of the following is the smallest?
   A. monocyte count
   B. eosinophil count
   C. erythrocyte count
   D. basophil count
   E. lymphocyte count

498. Which of the following signifies the time required for a pacemaker signal to spread from the SA node to the ventricular myocadium?
   A. QT interval
   B. ST segment
   C. PR interval
   D. QRS duration
   E. RR interval

499. The most powerful chemical influence on the regulation of alveolar ventilation is:
   A. the partial pressure of oxygen in systemic arterial blood
   B. the partial pressure of carbon dioxide in systemic venous blood
   C. the concentration of hydrogen ions in pulmonary arterial blood
   D. P_{CO_2} in systemic arterial blood
   E. two of the preceding

500. If the oxygen carrying capacity of the blood is 20 vol. % and each gram of hemoglobin transports 1.34 ml of oxygen, the hemoglobin concentration is:
   A. 13.4 gms/dl
   B. 14.9 gms/dl
   C. 26.8 gms/dl
   D. 0.067 gms/dl
   E. 19 gms/dl

501. The bicuspid and tricuspid valves are closed:
   A. while the ventricles are in diastole
   B. by the movement of blood from the atria to the ventricles
   C. when the ventricles are in systole
   D. while the ventricles are filling
   E. while the atria are contracting and emptying

_____502. Given the following data, compute capillary filtration pressure: HBP = 30 torr; PPOP = 28 torr; THP = 3 torr; TPOP = 1 torr.
   A.   8 torr
   B.   5 torr
   C.   3 torr
   D.   1 torr
   E.   zero

_____503. According to Dalton's law, if Gas A exerts a pressure of 500 torr and is a mixture of Gas B (15%), Gas C (70%), Gas D (10%) and Gas E, the partial pressure of Gas E is:
   A.   50 torr
   B.   75 torr
   C.   25 torr
   D.   350 torr
   E.   250 torr

_____504. Which of the following, if infused into an A⁻ recipient, would result in minor but not major agglutination?
   A.   plasma from B⁺ whole blood
   B.   plasma from A⁺ whole blood
   C.   serum from A⁺ whole blood
   D.   AB⁻ whole blood
   E.   more than one of the preceding

_____505. The rate of pulmonary diffusion of gases:
   A.   decreases as pulmonary membrane thickness decreases
   B.   increases during inspiration and decreases during expiration
   C.   decreases as the surface area of the pulmonary membrane decreases
   D.   is directly proportional to the solubility of the gas in the pulmonary membrane
   E.   two of the preceding

_____506. According to Poiseuille's Law:
   A.   blood flow through the artery is directly proportional to the pressure differential
   B.   blood flow through an artery is directly proportional to the viscosity of the blood
   C.   blood flow is directly proportional to the fourth power of the length of the blood vessel
   D.   blood flow is not influenced by the radius of the blood vessel
   E.   two of the preceding

507. The lymphocyte that produces antibodies is the:
   A. T_8 lymphocyte
   B. plasma cell
   C. T_4 lymphocyte
   D. memory cell
   E. suppressor cell

508. Which of the following helps rather than opposes inflation of the lungs?
   A. airway resistance
   B. surface tension of alveolar fluid
   C. elastic recoil of the lung
   D. elastic recoil of the chest wall
   E. positive intrapulmonic pressure

509. All of the following describe the antibody-mediated immune response *except*:
   A. B-lymphocytes and helper T_4 cells are involved in this type of response
   B. B-lymphocytes divide and differentiate into plasma cells and memory cells
   C. plasma cells secrete antibodies
   D. re-infection will activate the memory cells to develop into antibody-secreting plasma cells
   E. suppressor T_8 cells function to turn off the plasma cell activity

510. Which pressure is the greatest?
   A. atmospheric pressure
   B. intra-alveolar pressure during inspiration
   C. intra-alveolar pressure during expiration
   D. interpleural pressure during inspiration
   E. interpleural pressure during expiration

511. Which of the following is the sum of three primary lung volumes?
   A. EC
   B. FRC
   C. IC
   D. VC
   E. RV

512. During normal resting expiration:
   A. intrapulmonic pressure falls below atmospheric pressure
   B. intrapulmonic volume increases
   C. intrapulmonic pressure increases
   D. intrapulmonic volume decreases
   E. both answers (C) and (D) occur

_____513. One of the following is abnormally high. Which one?
  A.   WBC: 8000 per microliter
  B.   RBC: 5.3×10^6 per microliter
  C.   ESR: 6.0 mm/hr
  D.   Hematocrit: 60 percent
  E.   MCH: 30 μμg (micromicrograms)

_____514. Which of the following is the largest?
  A.   inspiratory reserve volume
  B.   tidal volume
  C.   expiratory reserve volume
  D.   residual volume
  E.   inspiratory capacity

_____515. One of the following statements is false. Which one?
  A.   Cardiac output depends upon venous return.
  B.   Pressure in systemic veins is determined by venous resistance.
  C.   If distensibility of the arterial system is diminished, pulse pressure will increase.
  D.   Stimulation of a vasoconstriction nerve to arterioles increases peripheral resistance, blood pressure, and blood flow in the capillaries being supplied by the arterioles.
  E.   Two of the preceding.

_____516. End-systolic volume (ESV):
  A.   increases when the ventricle is less compliant
  B.   is reduced by an increase in atrial pressure
  C.   is influenced by the value of arterial pressure at EDV
  D.   cannot vary
  E.   is the amount of blood ejected per beat

_____517. Capillary filtration pressure will increase if the following pressure(s) increase(s):
  A.   hydrostatic blood pressure (capillary)
  B.   plasma protein osmotic pressure (oncotic pressure)
  C.   tissue protein osmotic pressure
  D.   tissue hydrostatic pressure
  E.   two of the preceding

_____518. Compared to the systemic circulation, the pulmonary circulation:
  A.   has a lower pressure gradient from beginning to end
  B.   offers less resistance to the flow of blood
  C.   has a higher capillary filtration pressure
  D.   has a lower rate of blood flow through the circuit
  E.   two of the preceding

469

_____519.   Given the following data, compute heart rate:
SV = 70 ml/beat
EDV = 80 ml
ESV = 10 ml
CO = 5.6 L/min
HR = _____

A.   60 BPM
B.   70 BPM
C.   80 BPM
D.   90 BPM
E.   cannot be computed without more data

_____520.   Which of the following produces histamine?
A.   plasma cell
B. T_8 cell
C. T_4 cell
D.   erythrocyte
E.   basophil or mast cell

_____521.   If mean arterial pressure is 80 torr, diastolic pressure could be:
A.   85 torr
B.   90 torr
C.   60 torr
D.   87 torr
E.   91 torr

_____522.   Which of the following hematocrits would be considered normal for either an adult male or adult female?
A.   10%
B.   20%
C.   30%
D.   45%
E.   60%

_____523.   Given the following data, determine alveolar ventilation: ADS: 100 ml; RR: 15 cpm; VC: 3500 ml; RV: 1200 ml; TV: 400 ml.
A.   1300 ml
B.   300 ml/min
C.   500 ml/min
D.   3000 ml/min
E.   4.5 L/min

_____524. Systolic blood pressure is 140 torr. Diastolic pressure is 70 torr. Mean arterial blood pressure is:

- A. 78.3 torr
- B. 88.3 torr
- C. 93.3 torr
- D. 108.3 torr
- E. 55 torr

_____525. During Stage III of blood coagulation:

- A. PCF is produced
- B. fibrin is produced
- C. calcium ions are required
- D. thrombin is formed
- E. two of the preceding

_____526. All of the following are characteristic of the neutrophil *except*:

- A. are extremely phagocytic
- B. displays diapedesis
- C. displays chemotaxis
- D. principal role is defense against bacterial invasion
- E. plays a major role in hunting down and destroying immune complexes

_____527. The terms "intrinsic pathway" and "extrinsic pathway" refer to:

- A. the formation of carbonic acid from water and carbon dioxide
- B. carbon dioxide transport from tissues to the lungs
- C. conduction of impulses from SA node to AV node and AV node to ventricular muscle
- D. systemic and pulmonary circulation
- E. stage I of blood coagulation

_____528. The amount of air remaining in the lung at the end of a normal tidal expiration is called:

- A. expiratory reserve volume
- B. residual volume
- C. functional residual capacity
- D. maximal voluntary ventilation
- E. inspiratory reserve volume

_____529. The first heart sound is a result of:

- A. turbulence of blood in the atria
- B. turbulence of blood in the ventricle
- C. closing of semilunar valves
- D. closing of atrioventricular valves
- E. none of the preceding

_____530. Partial pressure of carbon dioxide is greatest in:
A. interstitial fluid
B. alveolar air
C. pulmonary arterial blood
D. intracellular fluid
E. pulmonary venous blood

_____531. Which of the following would you expect to increase during exercise as compared to rest?
A. inspiratory Reserve Volume
B. tidal Volume
C. expiratory Reserve Volume
D. residual Volume
E. answers (A), (B) and (C)

_____532. The function of surfactant is to:
A. increase surface area of the lung
B. reduce thickness of the pulmonary membrane
C. reduce alveolar surface tension
D. increase alveolar surface tension
E. reduce pulmonary compliance

_____533. The following cell secretes interleukins that stimulates T_8 cells to differentiate and clone:
A. B-memory cells
B. T_4 helper cell
C. T_4 inducer cell
D. T_8 cytotoxic cell
E. T_8 suppressor cell

_____534. A combined WBC count is 12000 cells per microliter, of which 700 cells per microliter are eosinophils. This suggests:
A. high combined count but low eosinophil count
B. parasitic infection or autoimmune disease
C. high combined count and high eosinophil count
D. normal combined and eosinophil counts
E. two of the preceding

_____535. One of the following statements is true. Which one?
A. Cardiac output does not depend upon venous return.
B. Pressure in systemic veins is determined by venous resistance.
C. If distensibility of the arterial system is diminished, pulse pressure will not change.
D. Stimulation of a vasoconstriction nerve to arterioles increases peripheral resistance, blood pressure, and blood flow in the capillaries being supplied by the arterioles.
E. Two of the preceding are true.

536. Consider the relationship between flow (F), pressure (P), and resistance (R) to the flow of blood through a blood vessel and pick the correct statement from below:
A. $P = FR$
B. $R = PF$
C. if flow remains constant and resistance to flow increases, then pressure falls
D. an increase in blood viscosity increases vascular resistance
E. if pressure increases, flow will also increase unless resistance decreases

537. Elevation of pressure in the vena cava or in the right atrium may result in:
A. reflex inhibition of heart rate
B. decreased cardiac output
C. an accelerated heart rate with decreased cardiac output
D. bradycardia
E. none of the preceding

538. Stroke volume will be increased by all of the following except:
A. increased atrial pressure
B. increased ventricular pressure
C. increased ventricular contractility
D. increased ventricular compliance, i.e., an increase in ease with which the ventricle can be stretched
E. increased heart rate

539. When is prothrombin activated?
A. during agglutination of blood
B. during Stage I of blood coagulation
C. during Stage II of blood coagulation
D. during Stage III of blood coagulation
E. during clot retraction in hemostasis

540. Relative to the exchange of gases between alveolar air and pulmonary blood, it is correct to state that:
A. a decrease in the lung surface area available for exchange affects with equal magnitude the removal of CO_2 and the uptake of O_2 in the pulmonary capillaries
B. the pressure gradient favor CO_2 removal from pulmonary blood is greater than the pressure gradient favoring O_2 uptake by the pulmonary blood
C. increasing alveolar ventilation will tend to decrease the CO_2 pressure gradient
D. decreasing alveolar ventilation will tend to decrease the O_2 pressure gradient
E. two of the preceding statements are correct

473

_____541.   The hormone that stimulates the bone marrow to produce more erythrocytes when tissue hypoxia occurs is called:
A.   interferon alpha
B.   erythrogenin
C.   erythropoietin
D.   lymphokine
E.   interleukin II

_____542.   Which of the following would you expect not to increase during exercise compared to rest?
A.   inspiratory reserve volume
B.   tidal volume
C.   expiratory reserve volume
D.   residual volume
E.   answers (A), (C), and (D)

_____543. T_4 lymphocytes:
A.   become cytotoxic lymphocytes
B.   engulf and destroy immune complexes
C.   suppress other types of lymphocytes
D.   induce T_8 lymphocytes to clone
E.   two of the preceding

_____544.   Systole is defined as:
A.   contraction of cardiac muscle
B.   contraction of ventricular muscle only
C.   contraction of atrial muscle only
D.   maximum arterial pressure
E.   none of the preceding

_____545.   According to the oxyhemoglobin dissociation curve:
A.   the higher the P_{CO_2} the higher the Hgb saturation
B.   the lower the P_{CO_2} the lower the Hgb saturation
C.   at normal alveolar PO_2 most of the Hgb is saturated (more than 50%)
D.   at normal resting interstitial PO_2 most of the Hgb is saturated (more than 50%)
E.   two of the preceding

_____546.   Relative to pulmonary circulation:
A.   blood flow is slowest in veins
B.   blood pressure is highest in arterioles
C.   most of the resistance to flow occurs in capillary beds
D.   blood pressure is higher in capillaries than in veins but blood flow is slower in capillaries than in veins
E.   none of the preceding

_____547. The Hering Bruer reflex:
  A.   results in an increase in alveolar ventilation in response to an increased arterial P_{CO_2}
  B.   limits inspiration
  C.   depresses alveolar ventilation when arterial P_{O_2} is normal
  D.   increases alveolar ventilation in response to a decreased arterial pH
  E.   two of the preceding

_____548. Antigen presentation is a primary function of which cell?
  A.   macrophage
  B.   eosinophil
  C. T_8 cell
  D. T_4 cell
  E.   basophil

_____549. One of the following statements is true. Which one?
  A.   Pulse pressure is inversely related to stroke volume.
  B.   Pulse pressure is directly proportional to heart rate.
  C.   Pulse pressure is directly proportional to peripheral resistance.
  D.   Pulse pressure = systolic pressure + diastolic pressure.
  E.   Pulse pressure is inversely related to arterial distensibility.

_____550. The following product is equal to mean arterial blood pressure:
  A.   ESV x EDV   D.   CO x HR
  B.   CO x TPR   E.   CO x SV
  C.   SV x HR

_____551. During the cardiac cycle, the aortic semilunar valve closes because:
  A.   papillary muscles contract
  B.   aortic pressure exceeds ventricular pressure
  C.   pulmonary trunk pressure exceeds right atrial pressure
  D.   right ventricular pressure exceeds right atrial pressure
  E.   aortic pressure exceeds pulmonary trunk pressure

_____552. Which of the following is abnormal?
  A.   hemoglobin content: 15 gms/dl
  B.   systemic arterial blood pH: 7.41
  C.   neutrophil count: 6000 per mm_3
  D.   stroke volume at rest: 30 ml
  E.   cardiac output of right ventricle at rest: 5 liter/min

_____553. Given the following data, compute capillary filtration pressure: HBP = 35 torr; PPOP = 30 torr; THP = 3 torr; TPOP = 6 torr.
  A.   8 torr   D.   1 torr
  B.   5 torr   E.   Zero
  C.   3 torr

_____554. According to Boyle's law:
A.   oxygen diffuses into pulmonary blood because of a higher partial pressure in the alveoli
B.   air enters the lung during inspiration because intrapulmonic pressure is less than atmospheric as lung volume increases
C.   systemic arterial blood carries more oxygen than pulmonary arterial blood
D.   carbon dioxide is more soluble than oxygen in body water
E.   PK = V

_____555. Numerically, which of the following is the largest?
A.   monocyte count
B.   eosinophil count
C.   erythrocyte count
D.   basophil count
E.   lymphocyte count

_____556. One molecule of hemoglobin is capable of binding with and carrying _____ molecules of oxygen.
A.   8   D.   4
B.   2   E.   16
C.   1

_____557. Which of the following increases during inspiration?
A.   interpleural pressure
B.   intrapulmonic pressure
C.   intrathoracic pressure
D.   intraalveolar pressure
E.   none of the above

_____558. Which of the following, if infused into an A+ recipient, would result in minor and major agglutination?
A.   whole blood A^+
B.   whole blood AB^-
C.   plasma from whole blood O^-
D.   serum from whole blood AB^+
E.   two of the preceding

_____559. The amount of air remaining in the lung at the end of a normal tidal expiration is called:
A.   expiratory reserve volume
B.   residual volume
C.   functional residual capacity
D.   vital capacity
E.   inspiratory reserve volume

_____560. T4 helper lymphocytes:
A. suppress the immune response by inhibiting inducer cells
B. are also known as natural killer cells
C. act as memory cells for antibody production and secretion
D. secrete interleukin 1 to activate neutrophils
E. assist cytotoxic and suppressor cells

_____561. Given the following data, compute capillary filtration pressure: HBP = 30 torr; PPOP = 28 torr; THP = 3 torr; TPOP = 4 torr.
A. 8 torr
B. 5 torr
C. 3 torr
D. 1 torr
E. Zero

_____562. If mean arterial pressure is 80 torr, and pulse pressure is 60 torr, diastolic pressure is probably:
A. 85 torr
B. 90 torr
C. 60 torr
D. 87 torr
E. 91 torr

_____563. On destruction of the red blood cells and metabolism of hemoglobin:
A. all components of Hb are excreted except iron which is conserved
B. all components of the molecule are conserved
C. all components of the molecule are excreted
D. iron and protein (amino acids) are conserved and the remainder is excreted as waste in the urine or feces
E. the globin is conserved and the heme is excreted

_____564. Mean arterial blood pressure is 90 torr and diastolic blood pressure is 60 torr; pulse pressure is probably:
A. 10 torr
B. 30 torr
C. 120 torr
D. 150 torr
E. 90 torr

_____565. The following cell recognizes an antigen presented in combination with MHC II:
A. cytotoxic T_8
B. suppressor T_8
C. plasma cell
D. helper T_4
E. two of the preceding

_____566. On a standard electrocardiogram, the distance from the beginning of one P wave to the beginning of the next P wave is 20 millimeters. The recorded heart rate is therefore:
- A. 60 BPM
- B. 75 BPM
- C. 85 BPM
- D. 120 BPM
- E. 100 BPM

_____567. Given the following, compute heart rate: EDV = 90 ml, ESV = 20 ml, CO = 4.9 L/min.
- A. 80 BPM
- B. 70 BPM
- C. 60 BPM
- D. 72 BPM
- E. 65 BPM

_____568. Which of the following contains b antibodies but not d antibodies?
- A. whole blood AB⁻
- B. serum from whole blood O⁺
- C. plasma from whole blood B⁻
- D. whole blood A⁺
- E. two of the preceding

_____569. Which of the following is the smallest?
- A. inspiratory reserve volume
- B. tidal volume
- C. expiratory reserve volume
- D. residual volume
- E. inspiratory capacity.

_____570. The blood cell which orchestrates cell-mediated and antibody-mediated immune responses is the:
- A. eosinophil
- B. basophil
- C. plasma cell
- D. helper T cell
- E. memory B cell

_____571. Given the following, compute capillary absorption pressure: hydrostatic blood pressure = 20 torr; plasma protein osmotic pressure = 30 torr; tissue hydrostatic pressure = 5 torr; tissue protein osmotic pressure = 5 torr.
- A. 5 torr
- B. 10 torr
- C. 15 torr
- D. 20 torr
- E. 25 torr

_____572. In pulmonary blood arriving at an alveolus:
   A.   bicarbonate is released into the alveolus
   B.   hydrogen ions are formed so that blood pH decreases
   C.   chloride ions move into the red cells (chloride shift)
   D.   blood pH tends to increase as carbon dioxide diffuses into the alveolus
   E.   two of the preceding

_____573. Partial pressure of oxygen is greatest in:
   A.   pulmonary arterial blood
   B.   alveolar air
   C.   pulmonary venous blood
   D.   intracellular fluid
   E.   atmosphere

MATCHING. CHOOSE THE BEST ANSWER. SOME ANSWERS MAY BE USED MORE THAN ONCE OR NOT AT ALL.

A. 7500 ml/min   B. 35%   C. 5%   D. 19 ml/dl   E. None of the preceding

_____574. normal adult female oxygen-carrying capacity of the blood

_____575. normal differential monocyte count, adult male

_____576. normal adult male MCHC

_____577. normal resting left ventricular cardiac output, adult male, SV = 70 ml

_____578. alveolar ventilation, resting unlabored breathing, adult female, tidal volume – 500 ml

_____579. normal differential lymphocyte count, adult female

_____580. normal adult male hemoglobin content of the blood

_____581. pulmonary ventilation (MRV), eupnea, resting, adult male

A. FEV$_{2.0}$
B. V.C.
C. TV
D. Alveolar Ventilation
E. E.D.V.

_____582. 5.2 L/min

_____583. 500 ml

_____584. 94%

_____585. 4.8 L

_____586. 75 ml

A. hematocrit
B. hemoglobin content
C. differential WBC: lymphocyte
D. oxygen-carrying capacity of hemoglobin
E. oxygen-carrying capacity of blood

_____587. 30%

_____588. 1.34 ml/gm

_____589. 16 gms/dl

_____590. 41%

_____591. 18 ml/dl

A.   M.C.H.
B.   Hematocrit
C.   Oxygen-carrying capacity
D.   Hemoglobin content
E.   Differential WBC: neutrophil

_____592.   19 ml/dl

_____593.   67%

_____594.   30 $\mu\mu$g

_____595.   17 gms./dl

_____596.   45%

A.   serum B$^-$
B.   serum O$_+$
C.   plasma A$^-$
D.   whole blood AB$^-$
E.   whole blood B$^-$

_____597.   This blood or blood component contains anti-A (a) but not anti-D (d) antibodies.

_____598.   This blood or blood component could be given to an A$^+$ recipient without minor agglutination but only if the donor had not been sensitized to the Rh antigen.

_____599.   An individual possessing this blood or blood component could donate whole blood to an A$^+$ recipient and neither major nor minor agglutination would occur.

_____600.   A person sensitized to the Rh antigen and possessing this whole blood would have antiA (a) and anti-D (d) antibodies in the serum.

_____601.   A person possessing this blood or blood component could not receive any of the following plasma without experiencing minor agglutination: A$^+$, A$^-$, B$^+$, B$^-$, O$^+$, O$^-$.

A. venules
B. arterioles
C. capillaries
D. veins
E. arteries

_____602. The velocity of blood is least in this type of vessel.

_____603. Most of what is termed "peripheral resistance" occurs here.

_____604. Systolic pressures are lowest in these blood vessels.

_____605. Most of these blood vessels have resting systolic pressure between 25 and 15 Torr.

IF THE MAGNITUDE OF THE ITEM IN COLUME A IS GREATER THAN THE MAGNITUDE OF THE ITEM IN COLUMN B, CIRCLE A. IF THE MAGNITUDE OF THE ITEM IN COLUMN B IS GREATER THAN THE MAGNITUDE OF THE ITEM IN COLUMN A, CIRCLE B. IF ITEMS A AND B HAVE THE SAME MAGNITUDE, CIRCLE C

<u>COLUMN A</u>	<u>COLUMN B</u>
A B C 606. systemic blood flow at rest (liters/min)	pulmonary blood flow at rest (liters/min)
A B C 607. resting pulmonary vascular mean arterial pressure	resting systemic vascular mean arterial pressure
A B C 608. pulmonary compliance near the end of maximal expiration	pulmonary compliance near the maximal inspiration
A B C 609. carbon dioxide content of systemic arterial blood	carbon dioxide content of pulmonary arterial blood
A B C 610. reflex stimulation of the inspiratory center by high systemic arterial [H^+]	reflex stimulation of the inspiratory center by high systemic arterial CO_2
A B C 611. resting pulmonary vascular resistance	resting systemic vascular resistance
A B C 612. pulmonary compliance near the end of maximal inspiration	pulmonary compliance near the beginning of tidal inspiration
A B C 613. carbon dioxide content of alveolar air	carbon dioxide content of pulmonary arterial blood

IF THE FOLLOWING VALUE IS NORMAL, CIRCLE B. IF BELOW NORMAL, CIRCLE C. IF ABOVE NORMAL, CIRCLE A. PRINT YOUR ANSWER IN THE BLANK. ASSUME MALE, ADULT.

		ABOVE	NORMAL	BELOW
_____614.	R.B.C. – 5.8 million/µL	A	B	C
_____615.	W.B.C. – 8.5 thousand/µL	A	B	C
_____616.	Hgb – 18 gms/dL	A	B	C
_____617.	Platelets – 20,000/µL	A	B	C
_____618.	HCT – 35%	A	B	C
_____619.	MCHC – 34%	A	B	C
_____620.	MCH – 20 µµg	A	B	C
_____621.	Platelets – 20 x 10^{-3}/µL	A	B	C
_____622.	R.B.C. – 5.5 x 10^{-6}/µL	A	B	C
_____623.	W.B.C. – 10 x 10^{-4}/µL	A	B	C

PHYSIOLOGY OUTLINES 19-27

MULTIPLE CHOICE. SELECT THE BEST ANSWER.

_____624.　Distension of the duodenum:
　　　A.　reflexly relaxes the cardiac sphincter
　　　B.　reflexly contracts the pyloric sphincter
　　　C.　reflexly stimulates gastric motility
　　　D.　reflexly inhibits jejunum motility
　　　E.　two of the preceding

_____625.　Assuming normal levels of blood glucose, where and how much of the filtered glucose is reabsorbed by the nephron?
　　　A.　50% proximal tubule, 50% distal tubule
　　　B.　100% distal tubule
　　　C.　5% limbs of Henle
　　　D.　70% proximal tubule
　　　E.　100% proximal tubule

_____626.　Which of the following is (are) not present in saliva?
　　　A.　mucin
　　　B.　water
　　　C.　amylase
　　　D.　trypsin
　　　E.　two of the preceding

_____627.　Which of the following is <u>secreted</u> by the renal tubular cells?
　　　A.　sodium
　　　B.　bicarbonate
　　　C.　potassium
　　　D.　amino acid
　　　E.　glucose

_____628.　How many calories per hour does the average adult's metabolic rate burn?
　　　A.　300
　　　B.　40
　　　C.　70
　　　D.　120
　　　E.　200

_____629. At a plasma concentration of 0.2 mg/ml, the rate of urinary excretion of substance P is 10 mg/min. Plasma clearance of substance P by the kidneys is therefore:
A.   2.0 mg/min
B.   50 mg/min
C.   50 mg/ml
D.   50 ml/min
E.   none of the above

_____630. Quantitatively the most significant process by which a human loses heat when the environmental temperature exceeds the body temperature is:
A.   convection
B.   evaporation
C.   radiation
D.   piloerection
E.   conduction

_____631. Physiologically, GFR is reduced by:
A.   vasoconstriction of the afferent arterioles
B.   vasodilation of the afferent arterioles
C.   decreasing plasma protein concentration
D.   vasoconstriction of efferent arterioles
E.   two of the above

_____632. Fertilization of the ovum normally occurs in the:
A.   vagina
B.   uterus
C.   abdominal cavity
D.   uterine tube
E.   cervix

_____633. Renal compensation for a respiratory acidosis includes:
A.   retention of ammonia
B.   increased formation of titratable acid in urine
C.   reduced reabsorption of filtered bicarbonate
D.   increased alveolar ventilation
E.   answers A and B

_____634. Motility of the colon is:
A.   stimulated by norepinephrine
B.   increased by gastric distension
C.   decreased by duodenal distension
D.   inhibited by the vagus
E.   two of the preceding

485

635. Which of the following is NOT an action of the placenta?
A. delivery of waste from fetus to mother
B. exchange of gases between fetus and mother
C. delivery of antibodies from mother to fetus
D. delivery of water and electrolytes from mother to fetus
E. delivery of nutrients from fetus to mother

636. Two gonadotropins work together to produce mature ova. They are:
A. LTH and STH
B. LH and ACTH
C. TSH and FSH
D. FSH and LH
E. LH and LTH

637. One of the following enzymes functions best when the pH of its environment is very low. Which one?
A. trypsin
B. amylase
C. carboxypeptidase
D. sucrase
E. pepsin

638. Which of the following is an endopeptidase?
A. pepsin
B. maltase
C. amylase
D. enterokinase
E. HCL

639. The rate of gastric emptying is:
A. primarily controlled by the amount of sugar in saliva
B. increased when fats enter the duodenum
C. inhibited by distension of the duodenum
D. inversely proportional to fluidity of chyme
E. two of the preceding

640. Gastrin:
A. is a stomach enzyme
B. stimulates pepsinogen secretion
C. is released by sympathetic stimulation
D. digests protein
E. buffers HCL

641. Acids in duodenal chyme promote the release of this hormone:
A. sodium bicarbonate
B. ACTH
C. gastrin
D. secretin
E. ptyalin

_____642. Thyroxin contains:
A. steroid
B. threonine
C. tyrosine
D. TSH
E. none of the above

_____643. Approximately what percentage of filtered water is usually reabsorbed by the distal nephron?
A. 70%   D. 5%
B. 75%   E. 25%
C. 100%

_____644. Glycosuria is not a complication of:
A. non-insulin dependent diabetes (NIDD)
B. adrenal diabetes
C. insulin-dependent diabetes (Type I)
D. diabetes insipidus
E. two of the preceding

_____645. All by itself, chronically elevated plasma bicarbonate (systemic arterial blood) suggests:
A. respiratory acidosis
B. nonrespiratory alkalosis
C. respiratory alkalosis
D. nonrespiratory acidosis
E. compensation for a nonrespiratory acid-base disorder

_____646. As ECF osmotic pressure rises above normal:
A. the kidneys increase urine formation
B. ADH release increases
C. cells begin to osmotically gain water
D. the distal nephron becomes impermeable to water
E. salt retention increases

_____647. Which of the following elevates hepatic portal blood glucose the most?
A. lactase   D. insulin
B. sucrase   E. amylase
C. maltase

_____648. In a hypothetical case of uncompensated nonrespiratory acidosis, which of the following could be true for systemic arterial blood?
A. partial pressure of carbon dioxide = 25 torr
B. bicarbonate concentration = 20 mEq/L
C. PH = 7.39
D. partial pressure of carbon dioxide = 60 torr
E. bicarbonate concentration = 24 mEq/L

_____649. Capsular pressure = 3 torr, glomerular hydrostatic pressure = 55 torr, capsular osmotic pressure = 0, and plasma protein osmotic pressure = 27 torr. Glomerular filtration pressure is:
A.   31 torr
B.   10 torr
C.   25 torr
D.   81 torr
E.   none of the above

_____650. The time between fertilization of the ovum and implantation of the fertilized ovum in the uterus is approximately:
A.   less than 24 hours
B.   2 days
C.   5 to 10 days
D.   21 days
E.   one month

_____651. The mucosal reflex involves:
A.   inhibition of gastric motility
B.   stimulation of duodenal secretion
C.   stimulation of salivary secretion
D.   inhibition of gastric secretion
E.   stimulation of colon motility

_____652. After exercise, humans continue to use oxygen at greater than rest levels. The difference between post-exercise and resting levels of O_2 uptake is known as:
A.   anaerobic glycolysis
B.   formation of lactate from pyruvate
C.   oxidative phosphorylation
D.   the citric acid cycle
E.   the oxygen debt

_____653. Which substance promotes the release of bile from the gallbladder?
A.   gastrin
B.   secretin
C.   norepinephrine
D.   cholecystokinin
E.   bicarbonate.

_____654. Which hormone's plasma concentration peaks (is at its highest) about a week after ovulation?
A.   estrogen
B.   progesterone
C.   LH
D.   FSH
E.   none of the preceding

_____655. An inhibitory hormone that engages a cell surface receptor may exert its effects on the target cell by way of:
  A.   increasing cyclic AMP formation
  B.   an inhibitory G-protein
  C.   stimulating Ca^{++} release
  D.   turning on one or more of the cell's genes
  E.   blocking a sodium channel

_____656. Which process in the kidney leads to generation of <u>new</u> bicarbonate to replenish depleted bicarbonate reserves?
  A.   excretion of ammonia
  B.   excretion of titratable acid
  C.   excretion of sodium chloride
  D.   reabsorption of urea
  E.   A and B

_____657. In comparing endocrine control and neural control, it is true to say that in general:
  A.   endocrine controls are much faster
  B.   neural controls are longer-lasting
  C.   neural controls are more specific or localized
  D.   the two systems have opposite effects on most target cells
  E.   most organs are controlled by one or the other of the two control systems but not both

_____658. An antipyretic:
  A.   acts on the hypothalamus to cause peripheral vasoconstriction
  B.   stimulates the hypothalamic heat—gain center
  C.   lowers the hypothalamic set-point for core body temperature
  D.   stimulates the release of TSH
  E.   does none of the above

_____659. ACE inhibitors reduce the:
  A.   distal tubular reabsorption of glucose
  B.   tubular transport maximum for glucose
  C.   plasma concentration of angiotensin II
  D.   gastric secretion of HCL
  E.   release of bile from the gallbladder

_____660. Given the following data, compute glomerular filtration pressure:
  A.   125 ml/min
  B.   10 mm Hg
  C.   15 mm Hg
  D.   20 mm Hg
  E.   25 mm Hg

_____661. Which of the following could be associated with a value of 30 torr for systemic arterial plasma carbon dioxide partial pressure?
   A.   gastric vomiting
   B.   compensated nonrespiratory acidosis
   C.   voluntary apnea
   D.   uncompensated nonrespiratory alkalosis
   E.   two of the above

_____662. The various process of the gastrointestinal system are directly associated with certain types of gastrointestinal cells. Which of the following association is NOT valid?
   A.   motility — esophageal smooth muscle cells
   B.   secretion — gallbladder epithelial cells
   C.   digestion — pancreatic acinar cells
   D.   absorption — small intestinal epithelial cells
   E.   absorption — gastric mucosal cells

_____663. Aspirin poisoning, such as might occur when a person chronically ingests too much for relief from arthritis, may cause:
   A.   uncompensated respiratory acidosis
   B.   nonrespiratory acidosis
   C.   nonrespiratory alkalosis
   D.   respiratory alkalosis without renal compensation
   E.   alkalemia

_____664. Pancreatic lipase:
   A.   digests peptides
   B.   requires secretin for release
   C.   splits triglyceride into monoglyceride and free fatty acids
   D.   forms chylomicrons
   E.   two of the above

_____665. Diuresis can result from:
   A.   hypersecretion of ADH
   B.   efferent arteriolar vasodilatation
   C.   afferent arteriolar vasoconstriction
   D.   filtration of large solutes that are not reabsorbed
   E.   two of the preceding

_____666. Sodium chloride and _____ are important in establishing and maintaining the cortico-medullary osmotic gradient in the kidney.
   A.   glucose
   B.   amino acids
   C.   potassium
   D.   urea
   E.   ammonia

_____667. A person with uncontrolled diabetes mellitus also has emphysema (pulmonary disease). Arterial blood values are:

$$pH = 7.21$$
$$P_{CO_2} = 60 \text{ torr}$$
$$Plasma[HCO_3^-] = 20 \text{ mEq/L}$$

This subject has:
A. nonrespiratory (metabolic) acidosis
B. mixed respiratory and nonrespiratory acidosis
C. mixed nonrespiratory (metabolic) and respiratory alkalosis
D. respiratory alkalosis
E. compensated nonrespiratory acidosis

_____668. Peptidases:
A. split fatty acids from glycerides
B. are secreted by gastric, intestinal, and pancreatic cells
C. break bonds between amino acids
D. are found in saliva
E. two of the preceding

_____669. The enterogastric reflex involves:
A. inhibition of gastric secretion
B. inhibition of colon motility
C. stimulation of gastric secretion
D. stimulation of pancreatic enzyme secretion
E. relaxation of gastric sphincters

_____670. Parathyroid hormone:
A. decreases ECF calcium levels
B. stimulates osteoblasts
C. is secreted by parafollicular cells
D. stimulates osteoclasts
E. two of the preceding

_____671. Type II diabetes mellitus:
A. is due to hyposecretion of glucagon
B. is always treated by insulin administration
C. is due to hypersecretion of insulin
D. involves a lack of insulin affect on target cells due to a change in insulin receptor function
E. two of the above

_____672. Gluconeogenesis:
  A. occurs in skeletal muscle when glycogen is depleted
  B. is stimulated by insulin in order to maintain blood sugar levels
  C. occurs in the liver primarily after ingesting a meal high in carbohydrate content
  D. involves the formation of glucose from non-carbohydrate sources by liver and kidney
  E. two of the preceding

_____673. Which of the following does not control the passage of food material through the alimentary canal?
  A. upper esophageal sphincter
  B. cardiac sphincter
  C. sphincter of oddi
  D. pyloric sphincter
  E. ileo-caecal sphincter

_____674. Glucagon:
  A. is a glucocorticoid
  B. stimulates hepatic glycogenolysis
  C. functions as an antagonist to hydrocortisone
  D. inhibits gluconeogenesis
  E. two of the preceding

_____675. Counter-current multiplication involves the:
  A. proximal and distal tubules
  B. distal and collecting tubules
  C. ascending and descending limbs of henle
  D. proximal tubule and vasa recta
  E. collecting tubule and vasa recta

_____676. In the normal kidney, the plasma clearance of insulin (a protein hormone) is:
  A. 125 ml/min
  B. 60 ml/min
  C. 250 ml/min
  D. zero
  E. none of the preceding

_____677. TSH - RH:
  A. release is accelerated when ECF osmotic pressure decreases below normal
  B. indirectly promotes elevation of plasma T_3
  C. stimulates cells of the neurophypophysis
  D. was the first hypothalamic inhibitory hormone to be discovered
  E. release is depressed when plasma iodine levels decrease below normal

_____678.　The juxtaglomerular apparatus:
- A. 　increases GFR when tubular transport maxima are exceeded
- B. 　reduces glomerular blood pressure when solute concentration in distal tubular filtrate increases
- C. 　is part of the renal sympathetic nervous system
- D. 　consists in part of modified AA smooth muscle cells that form the macula densa
- E. 　secretes angiotensin to maintain renal blood pressure

_____679.　Given the following data, compute BMR:
caloric equivalent of 1 liter of oxygen = 4.825
body surface area = 1.5 square meters
6 minute oxygen consumption (Corrected to STP) = 1500 ml
The answer is: _____ $Cal/M^2/Hr$ (rounded off)
- A. 　38
- B. 　48
- C. 　40
- D. 　42
- E. 　72

_____680.　When plasma calcium rises above normal, _____ secretion is stimulated:
- A. 　aldosterone
- B. 　calcitonin
- C. 　norepinephrine
- D. 　ADH
- E. 　parathyroid hormone

_____681.　Which of the following would tend to cause ADH secretion to increase?
- A. 　cellular hydration
- B. 　excessive drinking of water
- C. 　diabetes insipidus
- D. 　cellular dehydration
- E. 　none of the above

_____682.　Hyposecretion of _____ may lead to tetany and death.
- A. 　calcitonin
- B. 　insulin
- C. 　PTH
- D. 　LTH
- E. 　ACTH

_____683.　Atrial natriuretic peptide (ANP):
- A. 　increases systemic arterial blood pressure
- B. 　is an endopeptidase
- C. 　increases water permeability of the distal nephron
- D. 　inhibits the release of ADH
- E. 　increases sodium reabsorption

_____684. Hypersecretion of _____, if severe, may cause calcium phosphate to precipitate in tissues such as lung and heart causing death.
A. calcitonin   D. LTH
B. insulin   E. ACTH
C. PTH

_____685. Renin release is stimulated by:
A. increased blood pressure in afferent arterioles
B. increased effective arterial blood volume
C. increased sodium chloride transport by macula densa cells
D. stimulation of renal sympathetic nerves
E. an increase above normal plasma potassium level

_____686. Which of the following is the indicator for compensation in nonrespiratory alkalosis?
A. P_{CO_2} = 30 torr
B. HCO_3^- = 34 mEq/L
C. HCO_3^- = 20 mEq/L
D. P_{CO_2} = 50 torr
E. P_{CO_2} = 40 torr

_____687. One of the following hormones elevates the plasma concentration of sodium by increasing renal tubular reabsorption. Which one?
A. ANP   D. glucagon
B. ADH   E. aldosterone
C. STH

_____688. Which of the following is the indicator for compensation in respiratory acidosis?
A. P_{CO_2} = 50 torr
B. HCO_3^- = 34 mEq/L
C. HCO_3^- = 24 mEq/L
D. P_{CO_2} = 30 torr
E. HCO^-_3 = 18 mEq/L

_____689. Which of the following is not a homeotherm?
A. dog
B. cat
C. snake
D. horse
E. two of the preceding

494

_____690. During the menstrual cycle _____ inhibits the release of gonadotropin releasing hormone.
   A. ACTH
   B. estrogen
   C. prolactin release inhibiting hormone
   D. testosterone
   E. renin

_____691. Which of the following molecules plays a role in establishing the cortico-medullary osmotic gradient in the kidney?
   A. hydrogen
   B. glucose
   C. amino Acids
   D. sodium Chloride
   E. none of the preceding

_____692. Walking fast (5 MPH) results in an energy expenditure of about 650 calories per hour. Assuming all of the energy required came from burning fat, how far would a person have to walk fast to burn the fat in one slice of cheese (about 10 grams of fat)?
   A. 90 feet
   B. 0.6 mile
   C. About a mile and a half
   D. 130 feet
   E. A quarter mile

_____693. Which of the following is reabsorbed in nearly equal amounts by proximal and distal parts of the nephron?
   A. sodium
   B. calcium
   C. chloride
   D. glucose
   E. tyrosine

_____694. In a normal adult oral glucose tolerance test, blood sugar "peaks" about _____ after ingestion.
   A. 10-20 minutes
   B. 30-60 minutes
   C. 60-90 minutes
   D. 20-120 minutes
   E. 120-180 minutes

_____695. This hormone inhibits gonadotropin secretion. It is:
   A. FSH-RH   D. epinephrine
   B. MSH   E. thyroxin
   C. melatonin

_____696. Membrane hormone receptors may be associated with:
- A. B-proteins
- B. D-proteins
- C. G-proteins
- D. P-proteins
- E. R-proteins

_____697. Light falling on the retina inhibits the release of this hormone. The hormone is:
- A. melanocyte stimulating hormone
- B. norepinephrine
- C. melatonin
- D. parathormone
- E. insulin

_____698. The following data: $[HCO_3^-]= 18$ mEq/L, $P_{CO_2}= 38$ torr, pH $= 7.30$, is indicative of:
- A. compensated respiratory acidosis
- B. uncompensated nonrespiratory acidosis
- C. uncompensated respiratory alkalosis
- D. compensated nonrespiratory alkalosis
- E. compensated nonrespiratory acidosis

_____699. In the _____ mode of intercellular signaling a secretory cell releases a chemical messenger which then engages a receptor on another part of the same cell.
- A. neuroendocrine
- B. holocrine
- C. exocrine
- D. autocrine
- E. paracrine

_____700. Which hormone acts to decrease aldosterone secretion?
- A. angiotensin II
- B. angiotensin I
- C. atrial natriuretic peptide
- D. antidiuretic hormone
- E. cyclic AMP

_____701. STH - RH (HGH-RH) :
- A. is a gonadotropin
- B. is an anterior pituitary hormone controlling ovarian function
- C. is an abbreviation for growth hormone
- D. stimulates release of growth hormone
- E. is a hormone that regulates ECF osmotic pressure

_____702. Calcium is required for all but which one of the following?
A. maintenance of normal sodium permeability in nerves
B. blood clotting to occur
C. formation of visual pigment
D. secretion of certain proteins
E. serving as a second messenger to allow some hormones to operate

_____703. Triiodothyronine production and release is controlled directly by the:
A. hypothalamus
B. neurohypophysis
C. adenohypophysis
D. middle pituitary
E. parathyroids

_____704. From which part of the alimentary canal is the greatest amount of water absorbed?
A. stomach
B. esophagus
C. large intestine
D. small intestine
E. pharynx

_____705. Receptive relaxation reflex involves:
A. duodenum
B. colon
C. ileum
D. stomach
E. esophagus

_____706. Testosterone:
A. enlarges the vocal cords
B. thickens the skin
C. stimulates muscular development
D. causes bone growth
E. all of the above

_____707. Which of the following temperature regulating mechanisms is not controlled by the hypothalamic heat-gain center?
A. shivering
B. peripheral vasoconstriction
C. increased BMR
D. sweating
E. T_3 secretion

_____708.	With respect to normal values, during compensated respiratory acidosis, usually the:
A.	P_{CO_2} will be low, pH will be low
B.	pH will be high, P_{CO_2} will be high
C.	$[HCO_3^-]$ will be high, P_{CO_2} will be low
D.	P_{CO_2} will be high, $[HCO_3^-]$ will be high
E.	P_{CO_2} will be low, $[HCO_3^-]$ will be low.

_____709.	A 70 kilogram person at rest is consuming 0.25 liter of oxygen per minute. What is the metabolic rate in calories per hour?
A.	42.1
B.	15.0
C.	17.5
D.	38
E.	72.4

_____710.	Cholycystokinin (CCK):
A.	stimulates gastric peristalsis
B.	inhibits release of bile
C.	stimulates release of pancreatic hormones
D.	inhibits intestinal motility
E.	stimulates release of trypsinogen and chymotrypsinogen

_____711.	Which of the following is NOT characteristic of conditions prevailing for a BMR test?
A.	fasting for at least 12 hours
B.	person at complete rest for 30 to 60 minutes prior to test
C.	patient received an enema one to two hours before test
D.	surrounding temperature approximately 25 degrees centigrade
E.	usually in the morning after a restful sleep

_____712.	Obligatory water reabsorption:
A.	occurs in the proximal tubule
B.	occurs in the collecting tubule
C.	occurs deep in the renal medulla
D.	is not primarily controlled by ADH
E.	two of the preceding

_____713.	In respiratory acidosis:
A.	the kidneys compensate by conserving ammonia
B.	plasma bicarbonate levels will slowly fall
C.	systemic arterial P_{CO_2} will be low
D.	renal tubular cells will secrete more H^+ into urine
E.	two of the preceding

_____714. Capsular pressure = 10 torr, plasma protein osmotic pressure = 30 torr, glomerular hydrostatic pressure = 48 torr. The maximum that afferent arteriolar pressure could fall before glomerular filtration ceases is:
A. 6 torr
B. 7 torr
C. 12 torr
D. 9 torr
E. 10 torr

_____715. The blood concentration of _____ "peaks" last during the ovarian cycle of a normal 24 year old female:
A. estrogen
B. FSH
C. progesterone
D. LH
E. STH

_____716. The second messenger system may involve:
A. adenylate cyclase/cyclic AMP
B. phosphatidylinositol and diacylglycerol
C. receptor linked ion channels
D. both (A) and (B)
E. (A), (B), and (C)

_____717. Which of the following produces an increase in renal blood flow?
A. angiotension II
B. norepinephrine
C. glucagon
D. thromboxane
E. ANF

_____718. The corpus luteum is maintained during early pregnancy by:
A. progesterone
B. follicle stimulating hormone
C. human chorionic gonadotropin
D. estrogen
E. prolactin

_____719. Grave's disease involves:
A. hyposecretion of T_3
B. the formation of an antibody that stimulates T_3 release
C. hypersecretion of the adrenal medulla
D. myxedema as a complication
E. none of the preceding

499

_____720. Insulin:
- A. promotes cellular uptake of glucose
- B. controls ECF calcium levels
- C. is a hyperglycemic hormone
- D. is secreted by pancreatic alpha cells
- E. two of the preceding

_____721. The metabolic system is about _____ percent efficient.
- A. 10
- B. 20
- C. 33
- D. 50
- E. 66

_____722. The smallest end-product resulting from lipase activity is (are):
- A. ketoacids
- B. glycerol
- C. monosaccharides
- D. amino acids
- E. triglycerides

_____723. The cells involved in the formation and resorption of bones are the:
- A. osteoclasts
- B. osteoblasts
- C. osteocytes
- D. A and B
- E. A, B, and C

_____724. The enterogastric reflex involves:
- A. inhibition of duodenal peristalsis by the stomach
- B. stimulation of duodenal peristalsis by the stomach
- C. inhibition of duodenal peristalsis by the colon
- D. inhibition of gastric peristalsis by the duodenum
- E. stimulation of gastric peristalsis by the duodenum

_____725. A naked person standing in a room with an air temperature of 72EF, no circulation of air and no sunlight, will lose body heat primarily by:
- A. convection
- B. evaporation
- C. conduction
- D. radiation
- E. insensible perspiration

_____726. Approximately what percentage of filtered water is usually reabsorbed by the kidney tubules?
A. 1 percent
B. 20 percent
C. 15 percent
D. 70 percent
E. 99 percent

_____727. Which of the following is associated with an increase in the intracellular formation of cyclic AMP?
A. B- protein
B. D- protein
C. Gi protein
D. Gs protein
E. myosin

_____728. A decrease in serum calcium promotes, for corrective purposes:
A. release of calcitonin
B. hypersecretion of TRH
C. release of insulin
D. release of PTH
E. release of cortisone

_____729. How many calories per hour does the average adult's metabolic rate burn?
A. 300
B. 40
C. 70
D. 120
E. 2000

_____730. Midway in a 28-day menstrual cycle:
A. serum LH is at its peak (high concentration)
B. serum progesterone is at its peak
C. the ovary stops producing estrogens
D. the corpus albicans forms
E. two of the preceding

_____731. Secretin:
A. inhibits gastric secretion
B. stimulates gastric peristalsis
C. stimulates release of $NaHCO_3$ by pancreas
D. stimulates release of pancreatic insulin
E. is the same hormone as pancreozymin

_____732. The synthesis of T$_4$ requires:
- A. glycerol
- B. alanine
- C. tyrosine
- D. threonine
- E. fatty acid

_____733. The following data: $(HCO_3^-) = 24$ mEq/L, $P_{CO_2} = 40$ torr, pH = 7.41, is indicative of:
- A. compensated nonrespiratory acidosis
- B. compensated respiratory acidosis
- C. compensated respiratory alkalosis
- D. compensated nonrespiratory alkalosis
- E. none of the preceding

_____734. Which of the following are absorbed primarily into the intestinal lymphatics?
- A. monosaccharides
- B. disaccharides
- C. glycerides
- D. dipeptides
- E. amino acids

_____735. Which of the following splits fatty acids from triacylglycerol?
- A. amylase
- B. sucrase
- C. lipase
- D. ribonuclease
- E. lactase

_____736. Most of the ATP generated during the metabolism of one mole of glucose to CO_2, H_2O, and E is generated in the process of:
- A. anaerobic glycolysis
- B. conversion of pyruvate to acetyl COA
- C. the Kreb's Cycle (Tricarboxylic Acid Cycle)
- D. oxidative phosphorylation
- E. conversion of pyruvate to lactate

_____737. Which of the following has the greatest effect on increasing GFR?
- A. AA vasodilation, EA vasodilation
- B. EA vasodilation, AA vasoconstriction
- C. EA vasoconstriction, AA vasoconstriction
- D. AA vasodilation, EA vasoconstriction
- E. vasoconstriction of interlobular artery

_____738. Antidiuretic hormone (vasopressin):
  A. release is accelerated when ECF osmotic pressure falls below normal
  B. inhibits renin release
  C. is produced by the adenohypophysis
  D. release is inhibited by ANF
  E. decreases water permeability of the distal nephron

_____739. Determination of a male subject's basal metabolic rate by indirect calorimetry indicates an oxygen consumption of 1.2 liters for six minutes. (1 liter O_2 = 4.825 Cal.) His BSA = $2M^2$. His calculated basal metabolic rate is:
  A. higher than normal
  B. 72 Cal./M^2/hr
  C. compatible with hypothyroidism
  D. suggestive of TRH hypersecretion
  E. approximately 36 Cal.M^2/hr

_____740. The respiratory response to a renal-induced acidosis is to:
  A. excrete CO_2 at a rate equal to its production
  B. excrete less CO_2 than is being produced
  C. excrete more CO_2 than is being produced
  D. decrease alveolar ventilation
  E. two of the preceding

_____741. Given the following data: $[HCO_3^-]$= 24 mEq/L, P_{CO_2} = 20 torr, pH = 7.50 (systemic arterial blood); the most likely acid-base disorder is:
  A. compensated respiratory acidosis
  B. compensated nonrespiratory alkalosis
  C. compensated respiratory alkalosis
  D. compensated nonrespiratory acidosis
  E. uncompensated respiratory alkalosis

_____742. A decrease in serum sodium promotes, for corrective purposes:
  A. follicle cell release of T_4
  B. mineralocorticoid release
  C. release of insulin
  D. release of SRH
  E. two of the preceding

_____743. The energy liberated by catabolic processes can be converted into:
A. metabolic heat
B. external work
C. chemical storage
D. both A and B
E. A, B, and C

_____744. Cholecystokinin (CCK):
A. stimulates secretin release
B. stimulates HCO_3^- secretion
C. stimulates gastric peristalsis
D. stimulates salivary secretion
E. stimulates bile release

_____745. Gluconeogenesis:
A. occurs in skeletal muscle when glycogen is depleted
B. is stimulated by insulin in order to maintain blood sugar levels
C. occurs in the liver primarily after ingesting a meal high in carbohydrate content
D. involves the formation of glucose from non-carbohydrate sources by liver and kidney
E. two of the preceding

_____746. $[HCO_3^-]$= 15 mEq/L, P_{CO_2} = 20 torr, pH = 7.50. This data could result from:
A. renal failure to excrete ammonia
B. depression of medullary respiratory centers
C. aspirin poisoning
D. chronic diarrhea
E. anxiety and involuntary hyperventilation

_____747. Assuming 75°F environmental temperature on a windless, cloudy day, the principal means of non-evaporative heat loss by a person dressed in shorts and a T-shirt and walking is:
A. radiation
B. conduction
C. convection
D. sweating
E. piloerection

_____748. Pyretics:
A. act on the hypothalamus to cause peripheral vasodilatation
B. stimulate the hypothalamic heat-gain center
C. lower the hypothalamic set-point for core body temperature
D. are chemicals that reduce fever
E. inhibit the release of TSH

749. Glucocorticoids:
   A. are secreted by alpha cells
   B. are hypoglycemic hormones
   C. are controlled by GRH
   D. two of the above
   E. none of the above

750. The RENIN-ANGIOTENSIN mechanism that is important in solute regulation by the kidney involves:
   A. glucagon
   B. parathyroid hormone
   C. glucocorticoids
   D. mineralocorticoids
   E. oxytocin

751. The process by which glucose is synthesized from noncarbohydrate source is:
   A. glucogenesis
   B. glycogenesis
   C. gluconeogenesis
   D. glycogenolysis
   E. none of the preceding

752. Which fluid has the highest osmotic pressure?
   A. blood in afferent arteriole
   B. blood in efferent arteriole
   C. blood in renal artery
   D. blood in renal vein
   E. filtrate in proximal tubule

753. A fall in core body temperature below the hypothalamic set-point may result in an increased:
   A. thyroid release of T_3
   B. hypothalamic release of oxytocin
   C. adenohypophyseal release of SRH
   D. adenohypophyseal release of FSH
   E. secretion of enterogastrone

754. Estrogens (during the follicular phase of the ovarian cycle):
   A. stimulate the hypothalamus
   B. stimulate the anterior pituitary
   C. stimulate the ovaries
   D. all of the above

755. During anaerobic glycolysis:
   A.   oxygen is consumed and carbon dioxide is produced
   B.   4 moles of ATP are formed per mole of glucose metabolized
   C.   4 hydrogen atoms are produced
   D.   two of the above
   E.   none of the above

756. The ovarian cycle consists of the:
   A.   menstrual phase
   B.   luteal phase
   C.   follicular phase
   D.   both A and C
   E.   A, B, and C

757. This hormone secreted by the small intestinal mucosa stimulates the intercalated duct cells of the pancreas to secrete sodium bicarbonate.
   A.   Secretin   D.   Enterokinase
   B.   Cholecystokinin   E.   Enterogastrone
   C.   Gastrin

758. Atrial Natriuretic Factor:
   A.   acts to increase blood volume
   B.   acts to increase blood pressure
   C.   decreases glomerular permeability
   D.   increases renal excretion of salt and water
   E.   stimulates aldosterone secretion

759. The plasma clearance of substance X is 70 ml per minute at a plasma concentration of 0.2 mg of X per ml. The rate of urinary excretion of substance X is:
   A.   14 mg per min   D.   400 mg per min
   B.   40 mg per min   E.   160 mg per min
   C.   16 mg per min

760. The process of oxidative phosphorylation:
   A.   occurs within cytoplasm
   B.   generates hydrogen atoms
   C.   produces water
   D.   consumes 24 moles of ATP
   E.   two of the preceding

761. Glucagon:
   A.   is a glucocorticoid
   B.   stimulates hepatic glycogenolysis
   C.   functions as an antagonist to hydrocortisone
   D.   inhibits gluconeogenesis
   E.   two of the preceding

_____762.   Using data in the diagram below, compute filtration pressure.

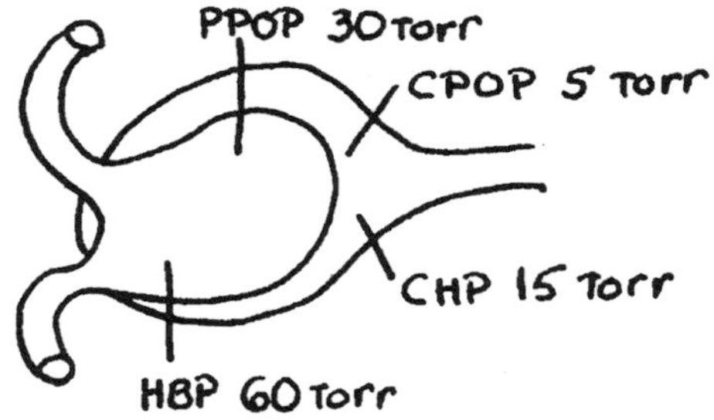

PPOP 30 Torr
CPOP 5 Torr
CHP 15 Torr
HBP 60 Torr

A.   5 torr
B.   10 torr
C.   15 torr
D.   20 torr
E.   Zero

_____763.   Which of the following molecules plays a role in establishing the cortico-medullary osmotic gradient in the kidney?
A.   urea
B.   glucose
C.   amino acids
D.   sodium chloride
E.   two of the preceding

_____764.   The various process of the gastrointestinal system are directly associated with certain types of gastrointestinal cells. Which of the following association is *NOT* valid?
A.   secretion—esophageal smooth muscle cells
B.   contraction—gallbladder smooth muscle cells
C.   digestion—pancreatic acinar cells
D.   absorption—small intestinal epithelial cells
E.   absorption—gastric mucosal cells

_____765.   Sodium chloride and _____ are important in establishing and maintaining the cortico-medullary osmotic gradient in the kidney.
A.   glucose
B.   amino acids
C.   potassium
D.   urea
E.   ammonia

_____766.   Adrenal diabetes:
A.   may occur as a complication of Cushing's disease
B.   can result from hyposecretion of the adrenal cortex
C.   involves renal loss of water but not sugar
D.   is more commonly known as Addison's disease
E.   is treated by administration of cortisone

767. Cushing's disease is characterized by:
   A. excessive retention of salt and water
   B. decreased resistances to stress because of inadequate glucocorticoid secretion
   C. lowered blood volume, blood pressure, and cardiac output
   D. excessive retention of potassium
   E. none of the preceding

768. Estrogens (during the follicular phase of the ovarian cycle):
   A. inhibit the hypothalamus
   B. inhibit the anterior pituitary
   C. inhibit the ovaries
   D. none of the above

769. Addison's disease is characterized by:
   A. excessive retention of salt and water
   B. elevated blood volume, blood pressure, and cardiac output
   C. increased levels of adrenocortical sex hormones
   D. depletion of extracellular sodium and water
   E. adrenal diabetes

770. Chronic exposure to ultraviolet light radiation from the sun stimulates the release of:
   A. MSH - RIH
   B. FSH
   C. MSH
   D. LTH
   E. LH

771. Which of the following contains the greatest amount of energy available for transfer to ATP?
   A. amino acid
   B. fatty acid
   C. glycerol
   D. glucose
   E. maltose

772. The time between fertilization of the ovum and implantation of the fertilized ovum in the uterus is approximately:
   A. 5 to 10 days   D. 2 days
   B. fewer than 24 hours   E. 9 months
   C. 3 weeks

773. Target cell surface receptors are utilized by which of the following:
   A. insulin   D. glucocorticoids
   B. androgens   E. mineralocorticoids
   C. estrogens

_____774. An arterial blood pH in excess of 7.45 resulting from a gain of strong base occurs during:
  A.   respiratory acidosis
  B.   respiratory alkalosis
  C.   metabolic acidosis
  D.   metabolic alkalosis
  E.   both B and D, but not A or C

_____775. In the normal kidney, the plasma clearance of insulin is:
  A.   125 ml/min
  B.   60 ml/min
  C.   250 ml/min
  D.   zero
  E.   none of the preceding

_____776. Relative to normal, if [HCO_3^-] is low, P_{CO_2} is low, and pH is low in systemic arterial blood, the most likely acid-base disorder is:
  A.   respiratory acidosis
  B.   nonrespiratory acidosis
  C.   nonrespiratory alkalosis
  D.   respiratory alkalosis
  E.   combined respiratory and nonrespiratory acidosis

_____777. pH influences:
  A.   rate of chemical reactions in the body
  B.   permeability of cell membranes
  C.   cell structure
  D.   both A and B
  E.   A, B, and C

_____778. When ambient air temperature rises above skin surface temperature, the following heat-loss factor becomes a heat-gain factor:
  A.   evaporation
  B.   radiation
  C.   convection
  D.   conduction (object cooler)
  E.   none of the above

_____779. In non-respiratory alkalosis of non-renal origin:
  A.   systemic arterial P_{CO_2} will increase
  B.   plasma bicarbonate will fall below normal range
  C.   renal excretion of titratable acid will be accelerated
  D.   urine will contain an increased amount of ammonia
  E.   none of the preceding will occur

509

_____780. Which of the following systemic arterial values suggest compensation in nonrespiratory alkalosis?
A. $[HCO_3^-] = 31$ mEq/L
B. $Pco_2 = 30$ mmHg
C. $Pco_2 = 40$ mmHg
D. $[HCO_3^-] = 20$ mEq/L
E. $Pco_2 = 50$ mmHg

_____781. The following hormone(s) stimulate(s) the secretion of testosterone and the development of secondary sex characteristics:
A. FSH
B. LH (ICSH)
C. LTH
D. ACTH
E. none of the preceding

_____782. The ovarian cycle consists of the:
A. menstrual phase
B. luteal phase
C. follicular phase
D. only (A) and (C)
E. (A), (B) and (C)

_____783. A significant reduction in renal arterial blood pressure results in:
A. nonrespiratory alkalosis
B. decreased renin release
C. glucosuria
D. increased formation of angiotensin II in pulmonary blood
E. none of the preceding

_____784. Respiratory compensation for a nonrespiratory alkalosis is often not complete because:
A. the lungs can't eliminate co_2 fast enough
B. maximum voluntary ventilation is about 110L/min
C. co_2 builds up in body fluids and stimulates respiratory centers
D. removal of co_2 from blood inhibits respiratory centers
E. the kidneys retain more bicarbonate

_____785. Potassium depletion can occur as a result of:
A. vomiting
B. diabetes
C. use of diuretics
D. diarrhea
E. all of the above

_____786. Which enzyme degrades protein to peptones, proteoses, and polypeptides?
A. trypsin
B. peptidase
C. chymotrypsin
D. pepsin
E. HCl

_____787. An animal that regulates its body core temperature within a narrow range of acceptable values is called:
A. a geotherm
B. a homeotherm
C. an isotherm
D. a poikilotherm
E. a thermal

_____788. Parathyroid hormone:
A. increases ECF calcium levels
B. stimulates osteoblasts
C. is secreted by parafollicular cells
D. stimulates osteoclasts
E. two of the preceding

_____789. The organ of greatest importance in providing the enzymes required for the digestion of food is the:
A. parotid gland
B. stomach
C. liver
D. pancreas
E. duodenum

_____790. The enterogastric reflex involves:
A. inhibition of gastric secretion
B. inhibition of colon motility
C. stimulation of gastric secretion
D. stimulation of pancreatic enzyme secretion
E. relaxation of gastric sphincters

_____791. The plasma clearance of substance X is 80 ml per minute at a plasma concentration of 0.2 mg of X per ml. The rate of urinary excretion of substance X is:
A. 20 mg per min   D. 400 mg per min
B. 40 mg per min   E. 160 mg per min
C. 16 mg per min

_____792. Diuresis can result from:
- A. hyposecretion of ADH
- B. efferent arteriolar vasodilatation
- C. afferent arteriolar vasoconstriction
- D. filtration of large solutes that are reabsorbed
- E. two of the preceding

_____793. The major change in the ion concentration between the plasma and the interstitial fluid is primarily because of difference in concentration of:
- A. sodium
- B. calcium
- C. protein
- D. chloride
- E. potassium

_____794. High (above normal) systemic arterial P_{CO_2} increases:
- A. bicarbonate reabsorption
- B. secretion of hydrogen ion into urine
- C. reabsorption of filtered acids in proximal tubule
- D. loss of glucose in urine
- E. two of the preceding

_____795. Calcium and phosphate serum levels are controlled by:
- A. parathryoid hormone
- B. thyroid hormones T_3 and T_4
- C. glucagon
- D. insulin
- E. cholecystokinin

_____796. Type II diabetes:
- A. is called insulin dependent diabetes mellitus
- B. results from a lack of ADH
- C. is often characterized by insulin resistance
- D. is most likely an autoimmune disease
- E. two of the preceding

_____797. Which of the following hormones does not influence metabolic rate?
- A. parathyroid hormone
- B. thyroxin
- C. epinephrine
- D. insulin
- E. growth hormone

_____798. Approximately what percentage of the filtered amino acids are usually reabsorbed by the kidney tubules?
   A.   1%
   B.   20%
   C.   70%
   D.   99%
   E.   100%

_____799. TSH - RH:
   A.   release is accelerated when core body temperature rises to 38EC
   B.   indirectly promotes elevation of plasma T_3
   C.   stimulates cells of the neurohypophysis
   D.   was the first hypothalamic inhibitory hormone to be discovered
   E.   release is depressed when plasma iodine levels decrease below normal

_____800. Physical stress:
   A.   increases CRH release
   B.   increases ACTH release
   C.   increases glucocorticoid release
   D.   increases release of epinephrine
   E.   all of the above

_____801. Pyrogens:
   A.   act on the hypothalamus to cause peripheral vasodilation
   B.   lower the hypothalamic set-point for core body temperature
   C.   are drugs like aspirin that reduce fever
   D.   stimulate the hypothalamic heat gain center
   E.   two of the preceding

_____802. None of the ATP generated during the metabolism of one mole of glucose to CO_2, H_2O, and E is generated in the process of:
   A.   anaerobic glycolysis
   B.   conversion of pyruvate to acetyl COA
   C.   the Kreb's Cycle (Tricarboxylic Acid Cycle)
   D.   oxidative phosphorylation
   E.   oxidizing hydrogen to form water

_____803. Normal digestion of fats results in the production of:
   A.   ketones and glycerol
   B.   monoglycerides
   C.   fatty acids
   D.   two of the preceding
   E.   all of the preceding

_____804. The renal response to a chronically elevated P_{CO_2} in renal blood is:
  A. increased excretion of bicarbonate
  B. increased excretion of ammonium
  C. decreased secretion of hydrogen ion
  D. decreased secretion of fixed acids
  E. vasoconstriction of the renal artery

_____805. One of the following is a gonadotropin that stimulates spermatogenesis. Which one?
  A. LTH
  B. TSH
  C. HGH
  D. LH
  E. FSH

_____806. Two hormones are required for the development of the ovarian follicle to maturity. They are:
  A. LTH, FSH
  B. LH, FSH
  C. FSH, TSH
  D. LH, TSH
  E. none of the preceding

_____807. The average adult sitting at rest burns up _____ calories per hour.
  A. 20
  B. 50
  C. 100
  D. 200
  E. 300

_____808. When plasma calcium falls below normal, _____ secretion is stimulated:
  A. aldosterone
  B. calcitonin
  C. norepinephrine
  D. ADH
  E. parathyroid hormone

_____809. Which of the following stimulates the release of ADH?
  A. high ECF osmotic pressure
  B. ANF
  C. T_4
  D. urea
  E. two of the preceding

514

_____810. A person with chronic diarrhea also has emphysema (pulmonary disease). Arterial blood values are:

pH = 7.21
P_{CO_2} = 60 torr
Plasma$[HCO_3^-]$ = 20 mEq/L

This subject has:
A. nonrespiratory (metabolic) acidosis
B. mixed respiratory and nonrespiratory acidosis
C. mixed nonrespiratory (metabolic) and respiratory alkalosis
D. respiratory alkalosis
E. none of the above

_____811. As the juxtaglomerular apparatus operates:
A. sodium reabsorption in the proximal tubule is reduced
B. glomerular filtration rate is reduced
C. water reabsorption by the ascending limb of Henle occurs
D. renal excretion of bicarbonate increases
E. two of the preceding

_____812. Concerning the assimilation of fat:
A. all end products of lipid digestion are absorbed only into the central lacteal
B. in the absence of bile, lipid digestion does not occur
C. lipids are absorbed faster by the stomach than by the small intestine
D. if lipid absorption is diminished, so is the absorption of vitamins A, D, E, and K
E. two of the preceding

_____813. If the proximal tubular glucose transport maximum is exceeded:
A. the efferent arteriole will vasoconstrict
B. glucose will appear in the urine
C. the glucose concentrations in renal arterial and renal venous blood will be equal
D. the afferent arteriole will vasodilate
E. two of the preceding

_____814. All by itself, chronically elevated carbon dioxide content of systemic arterial blood suggests:
A. nonrespiratory (metabolic) acidosis
B. nonrespiratory (metabolic) alkalosis
C. respiratory acidosis
D. respiratory alkalosis
E. hypocapnia

_____815. A hormone that stimulates secretion by the gastric parietal cell is:
  A.   gastrin
  B.   pepsin
  C.   gastric inhibitory peptide
  D.   secretin
  E.   calcitonin

_____816. Type II Diabetes Mellitus:
  A.   is due to hyposecretion of glucagon
  B.   is treated by insulin administration
  C.   is due to hypersecretion of insulin
  D.   involves a lack of insulin affect on target cells due to a change in insulin receptor function
  E.   two of the above

_____817. Quantitatively, the most significant tubular reabsorption occurs:
  A.   in the proximal nephron
  B.   in the distal nephron
  C.   in the loop of Henle
  D.   deep in the rental medulla
  E.   in the juxtaglomerular cells

_____818. With respect to thyroid function:
  A.   hypersecretion of T_3 in infancy may lead to cretinism
  B.   iodine is required for all thyroid hormones
  C.   myxedema is caused by hyposecretion
  D.   exophthalmus may result from hypersecretion
  E.   two of the preceding

_____819. The following hormone stimulates spermatogenesis in the seminiferous tubules:
  A.   FSH
  B.   LH
  C.   LTH
  D.   ACTH
  E.   none of the preceding

_____820. Mineralocorticoids:
  A.   increase release of renin
  B.   tend to elevate blood pressure and cardiac output
  C.   mobilize fat from storage areas
  D.   decrease renal absorption of chloride
  E.   none of the preceding

_____821. Diabetes:
- A. may result from lack of ADH
- B. may result from lack of glucagon
- C. may result from hypersecretion of insulin
- D. only two of the preceding
- E. Answers A, B, and C

_____822. Urea and _____ are important in establishing and maintaining the cortico-medullary osmotic gradient in the kidney.
- A. glucose
- B. amino acids
- C. potassium
- D. sodium chloride
- E. ammonia

_____823. Hypersecretion of HGH (also known as STH) after adulthood has been reached results in:
- A. cretinism
- B. acromegaly
- C. Addison's disease
- D. Cushing's disease
- E. diabetes insipidus

_____824. The process of digesting starch to release glucose is:
- A. glucogenesis
- B. glycogenesis
- C. gluconeogenesis
- D. glycogenolysis
- E. none of the preceding

_____825. Parathyroid hormone and a hormone secreted by the parafollicular cells of the thyroid gland are important in regulating extracellular fluid levels of:
- A. iodine
- B. K^+
- C. Na^+
- D. Ca^{++}
- E. T_3 and T_4

_____826. Diabetes:
- A. may result from hyposecretion of a hormone but not hypersecretion
- B. may result from hypersecretion of a hormone but not hyposecretion
- C. is generally characterized by polyuria (increase urine flow) no matter which endocrine gland is malfunctioning
- D. always involves malfunction of the pancreas
- E. two of the preceding

_____827.   Nearly all of the digestion of lipids takes place in the:
   A.   stomach
   B.   colon
   C.   duodenum and jejunum
   D.   rectum
   E.   oral cavity

_____828.   A gastrointestinal process that is *NOT* regulated is:
   A.   small intestinal absorption of glucose
   B.   gastric secretion of HCl
   C.   gastric emptying
   D.   secretion of pancreatic enzymes
   E.   swallowing

_____829.   Thyroid synthesis of T_3 and T_4 requires:
   A.   iodine and glycine
   B.   tyrosine and iodine
   C.   albumin
   D.   globulin
   E.   calcitonin

_____830.   Mineralocorticoids:
   A.   increase renal reabsorption of NaCl
   B.   decrease renal reabsorption of H_2O
   C.   inhibit renein
   D.   act as osmotic diuretics
   E.   are controlled primarily via TRH

_____831.   Diuresis can result from:
   A.   hypersecretion of ADH
   B.   efferent arteriolar vasodilation
   C.   afferent arteriolar vasoconstriction
   D.   filtration of large solutes that are not reabsorbed
   E.   two of the preceding

_____832.   Which of the following controls the passage of food material from the stomach?
   A.   upper esophageal sphincter
   B.   cardiac sphincter
   C.   sphincter of Oddi
   D.   pyloric sphincter
   E.   ileo-caecal sphincter

_____833. Which of the following contains the least amount of energy available for transfer to ATP?
A.   amino acid
B.   fatty acid
C.   glycerol
D.   glucose
E.   maltose

_____834. Chronic diarrhea in an otherwise normal child is likely to result in:
A.   compensated respiratory alkalosis
B.   compensated nonrespiratory acidosis
C.   uncompensated respiratory acidosis
D.   combined respiratory and nonrespiratory alkalosis
E.   none of the preceding

_____835. The organ of greatest importance in providing the enzymes required for the digestion of food is the:
A.   parotid salivary gland
B.   stomach
C.   pancreas
D.   liver
E.   cecum

_____836. Parathyroid hormone (PTH):
A.   increases osteoblast activity
B.   increases osteoclast activity
C.   increases renal excretion of Ca^{++}
D.   stimulates cell metabolism and increases BMR
E.   lowers ECF Ca^{++}

_____837. The plasma clearance of substance X is 80 ml/min at a plasma concentration of 0.4mg per ml. The rate of urinary excretion of substance X is:
A.   32 mg/min
B.   200 mg/min
C.   20 mg/min
D.   320 mg/min
E.   .05 mg/min

_____838. Counter-current multiplication involves the:
A.   proximal and distal tubules
B.   distal and collecting tubules
C.   ascending and descending limbs of Henle
D.   proximal tubule and vasa recta
E.   collecting tubule and vasa recta

839. The intestinal *digestion* of triacylglycerols is facilitated by the incorporation of these molecules into:
   A.   large oil droplets
   B.   micelles
   C.   chylomicra
   D.   small emulsified fat particles
   E.   membrane-bound vesicles

840. Capsular pressure = 12 torr, plasma protein osmotic pressure = 28 torr, glomerular hydrostatic pressure = 48 torr. The maximum that afferent arteriolar pressure could fall before glomerular filtration ceases is:
   A.   6 torr
   B.   7 torr
   C.   9 torr
   D.   10 torr
   E.   12 torr

841. Respiratory compensation for a non-respiratory acidosis is often not complete because:
   A.   excessive removal of CO_2 from the blood reduces medullary respiratory center stimulation
   B. CO_2 builds up in the blood causing acidemia
   C.   maximal voluntary ventilation is 110 L/min
   D.   high plasma P_{CO_2} increases renal excretion of H^+
   E.   low plasma P_{CO_2} decreases renal reabsorption of bicarbonate

842. The end-products of enzymatic digestion are two molecules of glucose. The enzyme is:
   A.   lipase
   B.   sucrase
   C.   lactase
   D.   maltase
   E.   peptidase

843. Distension of the distal small intestine relaxes this valve and distension of the proximal large intestine constricts it. The sphincter (valve) is the:
   A.   lower esophageal
   B.   upper esophageal
   C.   pyloric
   D.   ileocecal
   E.   external anal

844. Quantitatively the most significant organ of absorption is the:
   A.   stomach   D.   cecum
   B.   duodenum   E.   colon
   C.   jejunum

_____845. A fall in core body temperature:
   A.   inhibits the release of TSH-RH
   B.   causes peripheral vasodilation
   C.   stimulates the release of calorigenic hormones
   D.   decreases BMR
   E.   causes two of the above

_____846. Antipyretics:
   A.   act on the hypothalamus to cause peripheral vasoconstriction
   B.   stimulate the hypothalamic heat-gain center
   C.   elevate the hypothalamic set-point for core body temperatures
   D.   are chemicals that reduce fever
   E.   stimulate the release of TSH

_____847. An animal that regulates its core body temperatures within a broad range of acceptable values is called:
   A.   a geotherm
   B.   a homeotherm
   C.   an isotherm
   D.   a poikilotherm
   E.   a thermal

_____848. Secretin:
   A.   stimulates CCK release
   B.   stimulates HCO_3^- secretion
   C.   stimulates gastric peristalsis
   D.   stimulates salivary secretion
   E.   is secreted by the pancreas

MATCHING. CHOOSE THE BEST ANSWER. SOME ANSWERS MAY BE USED MORE THAN ONCE, OR NOT AT ALL.

A.   hyposecretion but not hypersecretion
B.   hypersecretion but not hyposecretion
C.   neither hypersecretion nor hyposecretion
D.   either hypersecretion or hyposecretion

_____849.   primary cause of cretinism

_____850.   primary cause of Cushing's disease

_____851.   primary cause of diabetes

_____852.   primary cause of myxedema

_____853.   primary cause of acromegaly

_____854.   primary cause of Addison's disease

A.   oxidative phosphorylation.
B.   tricarboxylic acid cycle.
C.   neither A nor B.
D.   both A and B.

_____855.   when one mole of glucose is metabolized, only 2 net ATP are formed during this part of the metabolism

_____856.   nearly all of the energy stored in glucose is transferred to ATP during this reaction sequence

_____857. CO_2 and H_2O are produced during this (these) process(es)

_____858.   anaerobic glycolysis precedes

A. cretinism
B. giantism
C. acromegaly
D. NIDDM
E. IDDM

_____859. immunologic destruction of beta cells

_____860. hyposecretion of T3 in early childhood

_____861. hypersecretion of HGH during adolescence

_____862. hypersecretion of HGH after epiphyseal closures are complete

_____863. hyperglycemia resulting from a decrease in insulin effectiveness on target cells

	Ph	Pco_2 mmHg	HCO_3^- mEq/L
A	7.55	25	24
B	7.21	60	23
C	7.34	60	31
D	7.53	50	40
E	7.29	30	14

_____864. compensated nonrespiratory acidosis

_____865. uncompensated respiratory acidosis

_____866. compensated respiratory acidosis

_____867. uncompensated respiratory alkalosis

_____868. compensated nonrespiratory alkalosis

A. anterior pituitary
B. thyroid
C. adrenal cortex
D. parathyroid
E. posterior pituitary

_____869. hyposecretion and cretinism

_____870. diabetes insipidus and hyposecretion

_____871. hyposecretion and tetany

_____872. hypersecretion and Cushing's disease

_____873. hyposecretion and Addison's disease

A. conversion of pyruvate to acetyl-CoA
B. tricarboxylic acid cycle
C. oxidative phosphorylation
D. anaerobic glycolysis
E. conversion of pyruvic acid to lactic acid

_____874. CO_2, H, and ATP are generated during this reaction sequence

_____875. CO_2 and H but not ATP are produced during this reaction sequence

_____876. quantitatively, the most significant ATP generating sequence of reactions

_____877. a cytoplasmic reaction sequence not requiring oxygen but nevertheless generating ATP